Fundamentals of Electronics 3

Fundamentals of Electronics 3

Discrete-time Signals and Systems, and Quantized Level Systems

Pierre Muret

ISTE

WILEY

First published 2018 in Great Britain and the United States by ISTE Ltd and John Wiley & Sons, Inc.

ISTE Ltd
27-37 St George's Road
London SW19 4EU
UK

www.iste.co.uk

John Wiley & Sons, Inc.
111 River Street
Hoboken, NJ 07030
USA

www.wiley.com

Library of Congress Control Number: 2018930834

British Library Cataloguing-in-Publication Data
A CIP record for this book is available from the British Library
ISBN 978-1-78630-183-3

Contents

Preface

Today, we can consider electronics to be a subject derived from the theoretical advances achieved during the 20th Century in areas comprising the modeling and conception of components, circuits, signals and systems, together with the tremendous development attained in integrated circuit technology. However, such development led to something of a knowledge diaspora that this work will attempt to contravene by collecting the general principles at the center of all electronic systems and components, together with the synthesis and analysis methods required to describe and understand these components and subcomponents. The work is divided into three volumes. Each volume follows one guiding principle from which various concepts flow. Accordingly, Volume 1 addresses the physics of semiconductor components and the consequences thereof, that is, the relations between component properties and electrical models. Volume 2 addresses continuous time systems, initially adopting a general approach in Chapter 1, followed by a review of the highly involved subject of quadripoles in Chapter 2. Volume 3 is devoted to discrete-time and/or quantized level systems. The former, also known as sampled systems, which can either be analog or digital, are studied in Chapter 1, while the latter, conversion systems, are addressed in Chapter 2. The chapter headings are indicated in the following general outline.

Each chapter is paired with exercises and detailed corrections, with two objectives. First, these exercises help illustrate the general principles addressed in the course, proposing new application layouts and showing how theory can be implemented to assess their properties. Second, the exercises act as extensions of the course, illustrating circuits that may have been described briefly, but whose properties have not been studied in detail. The first volume should be accessible to students with a scientific literacy corresponding to the first 2 years of university education, allowing them to acquire the level of understanding required for the third year of their electronics degree. The level of comprehension required for the following two volumes is that of students on a master's degree program or enrolled in engineering school.

In summary, electronics, as presented in this book, is an engineering science that concerns the modeling of components and systems from their physical properties to their established function, allowing for the transformation of electrical signals and information processing. Here, the various items are summarized along with their properties to help readers follow the broader direction of their organization and thereby avoid fragmentation and overlap. The representation of signals is treated in a balanced manner, which means that the spectral aspect is given its proper place; to do otherwise would have been outmoded and against the grain of modern electronics, since now a wide range of problems are initially addressed according to criteria concerning frequency response, bandwidth and signal spectrum modification. This should by no means overshadow the application of electrokinetic laws, which remains a necessary first step since electronics remains fundamentally concerned with electric circuits. Concepts related to radio-frequency circuits are not given special treatment here, but can be found in several chapters. Since the summary of logical circuits involves digital electronics and industrial computing, the part treated here is limited to logical functions that may be useful in binary numbers computing and elementary sequencing. The author hopes that this work contributes to a broad foundation for the analysis, modeling and synthesis of most active and passive circuits in electronics, giving readers a good start to begin the development and simulation of integrated circuits.

Outline

1) Volume 1: Electronic Components and Elementary Functions [MUR 17a].

i) Diodes and Applications

ii) Bipolar Transistors and Applications

iii) Field Effect Transistor and Applications

iv) Amplifiers, Comparators and Other Analog Circuits

2) Volume 2: Continuous-time Signals and Systems [MUR 17b].

i) Continuous-time Stationary Systems: General Properties, Feedback, Stability, Oscillators

ii) Continuous-time Linear and Stationary Systems: Two-port Networks, Filtering and Analog Filter Synthesis

3) Volume 3: Discrete-time Signals and Systems, and Quantized Level Systems.

i) Discrete-time Signals: Sampling, Filtering and Phase Control, Frequency control circuits

ii) Quantized Level Systems: Digital-to-analog and Analog-to-digital Conversions

Pierre MURET
February 2018

Introduction

This third volume covers signals and systems dealing with variables or quantities that are discrete or quantized. This leads to distinguishing two chapters: the first concerning the discrete-time case and the second that of discrete (or quantized) levels. The electronic circuits and applications implemented are of analog, digital or mixed nature, and some make use of both types of discretization. Similar to the previous volume, it is fundamental to explain the signals and their properties in detail as well as the basic circuits that transform these signals before considering the functions performed by more complex arrangements, which we will refer to as systems.

The first chapter begins with the study of discrete-time signals, obtained by sampling continuous-time signals, first by means of ideal sampling, then by actual sampling or by using interpolation. The use of the Fourier transform is essential and allows us to demonstrate, on the one hand, equivalences between a discrete variable in one domain and the periodic nature of the quantity depending on the dual variable in the other domain and, on the other hand, the fundamental theorem which determines the possibility to preserve (or not) all the information contained within a signal when shifting from continuous-time to discrete-time domains, called the sampling theorem or the Shannon theorem. Basic analog circuits are described. The other transforms, relevant in cases where discretization can be applied in both time and frequency domains, are also indicated since these are the ones that are used in practice. Next follows the study of the measurement of the time delay and of the phase shift between periodic signals in circuits comprising basic analog and logic functions, which are now widely used. Since this measurement is only achieved once per period, the measured time and phase shifts become discrete quantities. However, in many cases, the approximation which consists of only considering the continuous-time domain, obtained by interpolation and assuming that stationarity is preserved, makes it possible to detail the operation of the analog phase-locked loop (PLL) and the correction strategies of this loop system. This approximation is also a means to establish a relation between phase and frequency, which proves very useful for the applications subsequently addressed. The PLL has undergone overly significant development since the 1970s, because it has allowed transformations of signals and their properties, which were very difficult or impossible to achieve without it, namely in areas such as instrumentation, computer sciences and communications (wireless broadcasting, wireline transmission, etc.) destined for conveying information. The main functions, grouped under the term "frequency synthesis," are described. Digital PLLs are also covered in detail.

The last part of this chapter is dedicated to samples systems analysed with the z - transform (ZT), just as with the Laplace transform in the case of continuous-time systems. The properties of the ZT are carefully presented in order to provide all the tools that will be used in the end of the first chapter and in the next chapter, including the new meaning for the plane of the complex z variable. The study of switched-capacitor circuits is then discussed in a didactic manner, because it is based on the principles of electrostatics which are simple but nonetheless not necessarily familiar to the readers when applied to capacitor networks. These circuits have experienced major developments because of the possibility to integrate them naturally within CMOS technology and they constitute the basic building blocks for analog sampled filters, and modern digital-to-analog conversion (DAC) or analog-to-digital conversion (ADC). The first chapter logically proceeds through the study of two types of sampled filters (with infinite impulse response [IIR] and with finite impulse response [FIR]) and their properties, as well as approximations useful to recover second order transfer functions in the frequency domain. The notion of transmittance in the plane of the z variable is developed for all the basic functions useful for building these filters. The synthesis methods of these filters are briefly described in order to introduce the readers to the use of numerical functions available in MATLAB® or SciLab software. On the one hand, it should be noted that FIR filters allow one to access properties inaccessible to IIR and analog filters and, on the other hand, that all processing and analyses based on the ZT can be applied without the need for specifying if the technology being used for the implementation is either analog or digital. In the first case, these are switched-capacitor circuits that were previously studied and that are used, while the basic principle of the numerical functions necessary for the second case is described to conclude this chapter. Finally, we proceed with showing the power of state variable analysis in the discrete-time domain and in the plane of the z variable for sampled systems. In effect, it provides direct access to the mathematical modeling of these systems characterized by their fundamental parameters, namely transmittance poles in the plane of the z variable. Provided that the computation of successive samples is performed by computerized means, it enables, in addition, the avoidance of all the approximations previously employed. This model paves the way for the exact computation of the sampled time response in the case of nonlinear systems and/or undergoing frequency variations strong enough so that certain parameters of their transfer characteristics depend thereupon, which is the case in PLLs.

In the second chapter, we consider the principles and implementations of systems dealing with quantized signals, as is the case for ADCs and DACs. The digital quantity is a number encoded onto n bits in the binary system. The quantization of a signal induces some degradation, the first of which being quantization noise, which is presented and analyzed. The other imperfections, which can be likened to errors disturbing the original signal after its conversion, are then connected to the electrical characteristics of these converters. DAC is detailed through the various principles that can be implemented, on the one hand, with resistor ladder networks, historically the first ones to have been used, and, on the other hand, with switched-capacitor circuits, well-adapted to CMOS technology. The reverse conversion is then presented along with its different possible principles, which all have in common the development of an approximation of the analog quantity in digital form, then reconverted and compared to the original quantity, increasingly more precise during the successive stages of the conversion. Looped systems are therefore the main subject. In general, the complexity of systems increases if it is desirable to reduce the conversion time, and the quality of the analog comparator (or analog comparators) determines a very significant part of the accuracy of the conversion.

Finally, “sigma-delta” or “delta-sigma” conversions are addressed, which are the most recent in the field. In its basic principle, the “delta-sigma” conversion is easily understood if deduced from that using a ramp voltage and a count. However, when it is desirable to increase the performance of this type of converter, one is confronted, on the one hand, with a significant sophistication of the modeling, especially in optimizing the signal-to-noise ratio and, on the other hand, with the stability problems of the loop because it is necessary to increase the order of the filter(s) beyond two. In order to solve these problems, a large number of concepts presented in the first chapter are utilized. The core of this type of converter is formed by the modulator, a closed system that processes signals with a significantly much lower number of bits than the initial or final number, desired or imposed, but at a much faster rate than that of the input or the output. The operations carried out by the modulator yield a loss of resolution that will be recovered later by the decimator filter, and a displacement of the noise spectrum toward higher frequencies where it will be more easily filtered. The first-order modulator is examined in the first place and then followed by a generalization to higher-order modulators, which makes it possible to establish the transfer functions for the signal and for the quantization noise. Several types of stable modulator are examined. The role and the way to build the decimator filter are discussed and, based on this analysis, a scaling of the different frequencies to be used can be proposed. Finally, the principle and the implementation of the digital-to-analog "delta-sigma" converter are presented. Although in theory it is deduced from the ADC by swapping digital and analog functions, it is preferable to describe it by proceeding to digital resolution and rate conversions before the final DAC and the associated filtering, which conforms to the practical implementation.

This entire volume thus presents discrete-time and quantized-level signals and systems, the transformations of these signals into continuous-time or continuous-level signals as well as reverse transformations, analog, digital or mixed circuits, effective to achieve these operations and the models capable of calculating, predicting and scaling the responses of these systems. Corrected exercises are provided in order to address specific cases not fully detailed in the course, in order to illustrate it, to complete it and to show the methods adapted to solve the presented problems.

1

Discrete-time Signals and Systems

1.1. Discrete-time signals

In the discretized time domain, where only specific moments are taken into consideration and identified, signals are represented by series of samples.

1.1.1. *"Dirac comb" and series of samples*

The "Dirac comb" distribution is the basic tool in the discrete-time domain.

1.1.1.1. *Dirac comb in the phase space and in the time domain*

Signals can be expressed by means of linear combinations of complex exponentials $\exp(jn\theta)$ in the time and frequency domains (see Chapter 1 of Volume 2 [MUR 17b]), n being an integer and θ an angle proportional to the angular frequency-time product. In the case of a linear combination of $2N+1$ terms having the same amplitude, n varying from $-N$ to $+N$, $I_N(\theta)=\sum_{n=-N}^{N}\exp(jn\theta)$, a function $I_N(\theta)$ is obtained which gives very sharp maxima (or lines) every time $\theta = 2k\pi$ (Figure 1.1) with k integer, because it is the only case in which the images of $\exp(jn\theta)$ in the complex plane are colinear and add up while their sum tends to cancel out when $\theta \neq 2k\pi$:

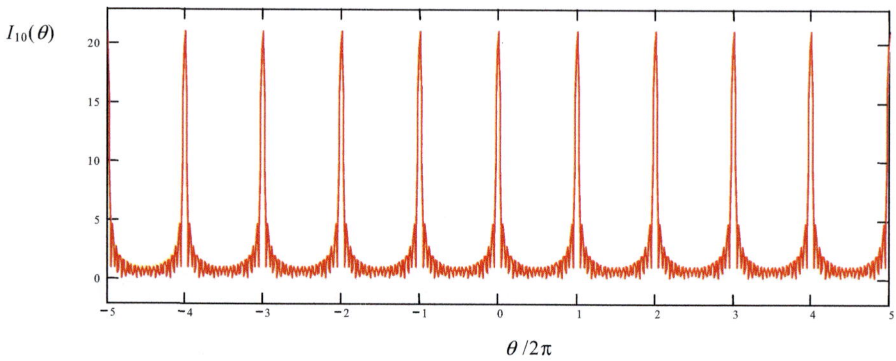

Figure 1.1. *Function $I_N(\theta)$ with N = 10*

The "Dirac comb" is a series of periodic Dirac impulses and can be defined from $I_N(\theta)$ by taking the limit for $N \to \infty$.

Provided that $\lim_{T_0 \to \infty} \left[2 f_0 \dfrac{\sin(2\pi f_0 t)}{2\pi f_0 t} \right] \to \delta(t)$ and $\lim_{T_0 \to \infty} \left[2 T_0 \dfrac{\sin(2\pi T_0 f)}{2\pi T_0 t} \right] \to \delta(f)$ (see Chapter 1 and the Appendix in Volume 2 [MUR 17b]), or more generally $\lim_{n \to \infty} \left[n \dfrac{\sin(n\theta)}{n\theta} \right] \to \delta(\theta)$, we define the "Dirac comb" distribution by means of a sum of Dirac impulses (Figure 1.2):

$$\sum_{k=-\infty}^{\infty} \delta\left(\frac{\theta}{2\pi} - k \right) = \lim_{N \to \infty} \sum_{n=-N}^{N} \exp(jn\theta) = \lim_{N \to \infty} \frac{\sin\left(\dfrac{2N+1}{2} \theta \right)}{\sin\left(\dfrac{\theta}{2} \right)}$$

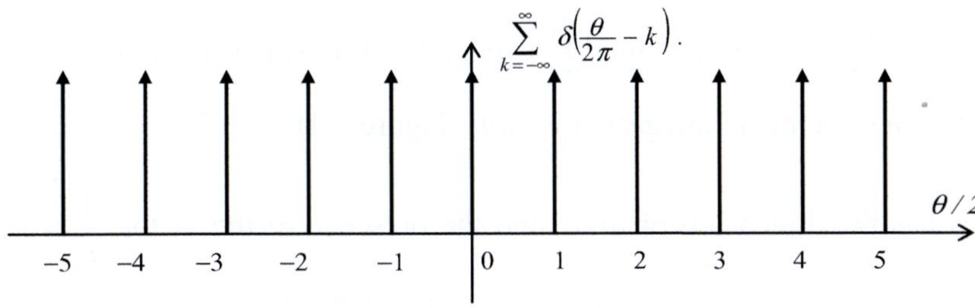

Figure 1.2. *Dirac comb in the phase space*

If $\theta = 2\pi f_0 t$, where $\frac{1}{T_0} = f_0$ is a fixed frequency and t is the time variable, we obtain, by changing variable in the distribution $\Sigma\delta_k$, a Dirac comb in the time domain $T_0 \Sigma\delta_{kT_0} = T_0 \sum_{k=-\infty}^{\infty} \delta(t - kT_0) = \lim_{N\to\infty} \sum_{n=-N}^{N} \exp(j2\pi n f_0 t)$ with a single line whenever $f_0 t = k$ (k being an integer), that is $t = k\ T_0$ (Figure 1.3).

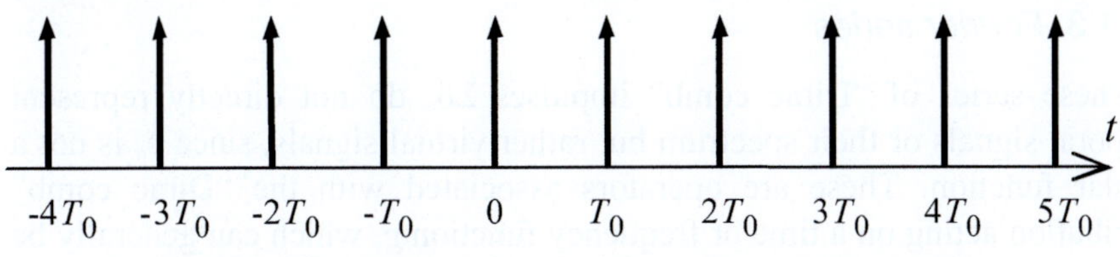

Figure 1.3. *Time "Dirac comb" distribution of time period T_0*

$T_0 \Sigma\delta_{kT_0}$ is dimensionless and $\Sigma\delta_{kT_0}$ has the dimension of a frequency, in order to preserve the dimension of the function to which the distribution is applied.

1.1.1.2. *Frequency Dirac comb*

Alternatively, if $\theta = 2\pi f T_0$, where $\frac{1}{f_0} = T_0$ is a fixed period and f is the frequency variable, we have a Dirac comb in the frequency domain

$f_0 \Sigma\delta_{nf_0} = f_0 \sum_{n=-\infty}^{\infty} \delta(f - nf_0) = \lim_{N\to\infty} \sum_{k=-N}^{N} \exp(j2\pi k f T_0)$ with a single line every time $fT_0 = n$ (n being an integer) or $f = n f_0$ (Figure 1.4).

$\Sigma\delta_{nf_0}$ has the dimension of time, and the presence of the factor $\frac{1}{T_0} = f_0$ can be verified by calculating the coefficient of the Fourier series of the frequency Dirac comb, which is periodic.

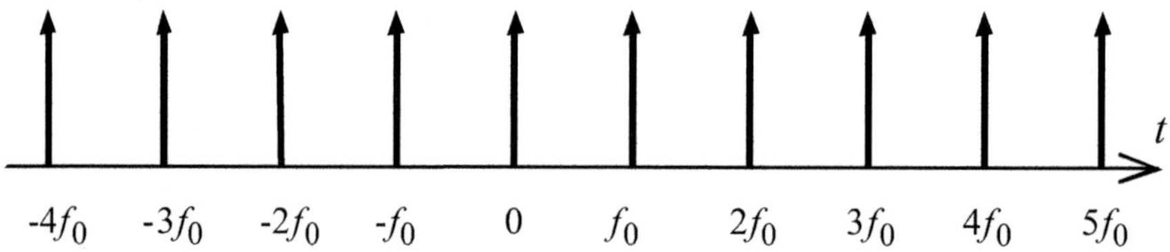

Figure 1.4. *Frequency "Dirac comb" distribution of frequency period f_0*

1.1.1.3. *Fourier series*

These series of "Dirac comb" impulses $\Sigma\delta_n$ do not directly represent temporal signals or their spectrum but rather virtual signals, since δ_n is not a regular function. These are operators associated with the "Dirac comb" distribution acting on a time or frequency function g, which can generally be denoted as a functional associated to $\Sigma\delta_n$: $< T_{\Sigma\delta n} , g >$ (see Appendix in Volume 2 [MUR 17b]). We can also consider them as bases of the vector spaces of series or discrete functions upon which the continuous-time signal or the continuous-frequency spectrum is projected by computing either the distribution applied to the signal or to the spectrum, or the distribution applied to the product of the signal or spectrum by exp ($\pm j2\pi ft$) when looking for the Fourier transform (FT) or its inverse.

The use of the "frequency Dirac comb" distribution corresponds to the decomposition of a periodic signal into Fourier series, already demonstrated in Chapter 1 of Volume 2 [MUR 17b] and derived from a description of the periodic time signals built on exponential function bases.

As a matter of fact, by taking the FT of a periodic signal $y_{T0}(t) = y_{T0}(t + kT_0)$ in the time domain, then reducing the interval to the period T_0 and finally by performing a summation over all periods, we find:

$$\mathrm{TF}\left[y_{T0}(t)\right]=\int_{-\infty}^{\infty} y_{T0}(t)\exp\left(-j2\pi f t\right)dt=\sum_{k=-\infty}^{\infty}\int_{-T_0/2}^{T_0/2} y_{T0}(t)\exp\left[-j2\pi f(t+kT_0)\right]dt$$

$$=\left[\int_{-T_0/2}^{T_0/2} y_{T0}(t)\exp\left(-j2\pi f t\right)dt\right]\sum_{k=-\infty}^{\infty}\exp\left(-j2\pi f k T_0\right)$$

expression of the product of the Dirac comb $f_0\sum_{n=-\infty}^{\infty}\delta(f-nf_0)=\lim_{N\to\infty}\sum_{k=-N}^{N}\exp(j2\pi k f T_0)$ by the integral between brackets, to be evaluated for the only line frequencies $f = n f_0 = \dfrac{n}{T_0}$. It is therefore possible to place the result of the integral calculated at $f = n f_0$ inside the summation by replacing f by $n f_0$ and write:

$$\mathrm{TF}\left[y_{T_0}(t)\right]=\sum_{n=-\infty}^{\infty} c_n\,\delta(f-nf_0) \text{ with the complex Fourier coefficient}$$

$$c_n=\frac{1}{T_0}\int_{-T_0/2}^{T_0/2} y_{T0}(t)\exp\left[-j2\pi\, n f_0 t\right]dt\,.$$

The inverse FT again yields the sum of the complex Fourier series, calculated by the distribution associated with the frequency Dirac comb applied to $\exp(j2\pi ft)$:

$$y_{T_0}(t)=\int_{-\infty}^{\infty}\sum_{n=-\infty}^{\infty} c_n\,\delta(f-nf_0)\exp\left(j2\pi f t\right)df=\sum_{n=-\infty}^{\infty} c_n\exp\left(j2\pi n f_0 t\right).$$

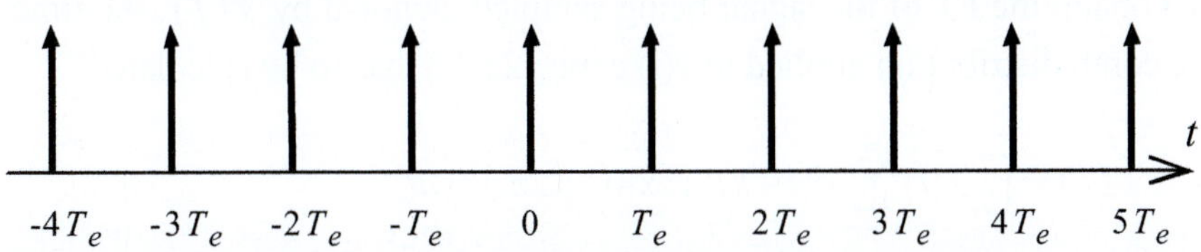

Figure 1.5. *Time "Dirac comb" distribution of period T_e*

In the case of the time Dirac comb, a (virtual) periodic signal of period T_e (Figure 1.5), the coefficients c_n are obtained by means of $\frac{1}{T_e}\int_{-Te/2}^{Te/2} \delta(t)\exp[-j2\pi n f_e t]dt = \frac{1}{T_e}$, which is a result independent of n. Consequently, the spectrum of the time Dirac comb (Figure 1.6) is also a Dirac comb, but in the frequency domain:

$$\frac{1}{T_e}\sum_{n=-\infty}^{\infty}\delta(f-nf_e)=f_e\sum_{n=-\infty}^{\infty}\delta(f-nf_e).$$

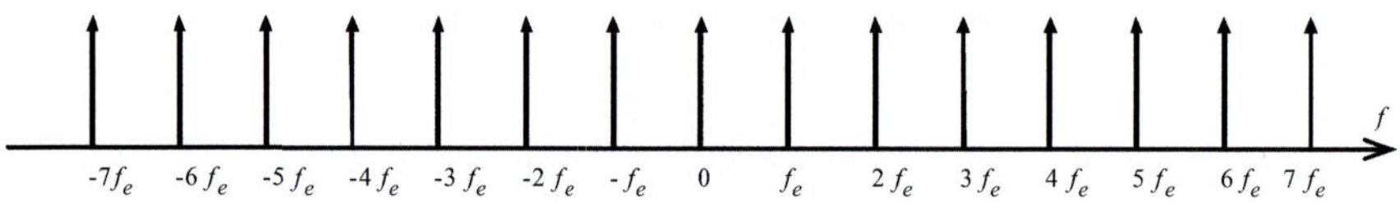

Figure 1.6. *Spectrum of the time Dirac comb (frequency Dirac comb)*

1.1.1.4. *Sampled (or discrete time) signal and periodicity of the spectrum*

The time Dirac comb corresponds to a base upon which a continuous time signal $y(t)$ can be projected to obtain a discrete signal or sampled regularly with the period $T_e = 1 / f_e$ in the time domain. To obtain any sample $y(kT_e)$, we calculate the distribution associated with a Dirac impulse $\delta(t-kT_e)$ applied to $y(t)$:

$$\int_{-\infty}^{\infty} y(t)\delta(t-kT_e)\,dt = y(kT_e).$$

To obtain the FT of the signal being sampled, denoted by $\hat{Y}(f)$, the time Dirac comb distribution applied to $y(t)\exp(-j2\pi f t)$ has to be calculated:

$$\hat{Y}(f)=\int_{-\infty}^{\infty} y(t)\sum_{k=-\infty}^{\infty}\delta(t-kT_e)\exp(-j2\pi f t)\,dt,$$

that is to say, the FT of the regular product $y(t)\times\left[\sum_{k=-\infty}^{\infty}\delta(t-kT_e)\right]$. It is thus the convolution product of the spectrum $Y(f)$ of $y(t)$ by the FT of the time Dirac comb, which is the frequency Dirac comb $f_e\sum_{n=-\infty}^{\infty}\delta(f-nf_e)$, namely:

$$\hat{Y}(f)=f_e\int_{-\infty}^{+\infty}\sum_{n=-\infty}^{+\infty}\delta(\upsilon-nf_e)\,Y(f-\upsilon)\delta\upsilon=f_e\sum_{n=-\infty}^{+\infty}Y(f-nf_e)$$

that is to say, the spectrum Y(f) of the continuous-time signal is duplicated periodically around the frequencies nf_e which are multiples of the sampling frequency, and is multiplied by f_e.

The spectrum of a sampled signal is therefore periodic (frequency domain, Figure 1.7):

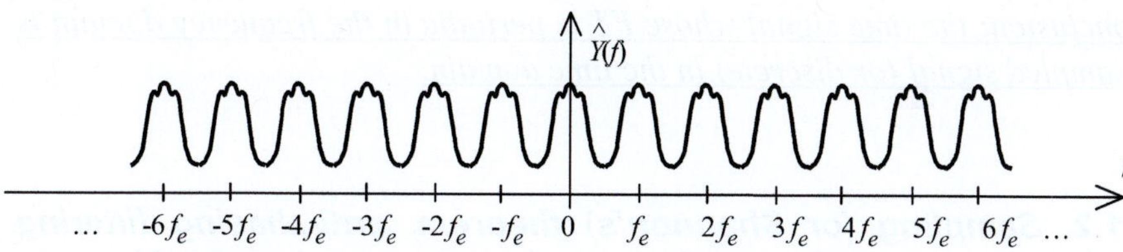

Figure 1.7. *Spectrum of a signal sampled at frequency f_e*

Conversely, it is possible to calculate the inverse FT of a periodic spectrum $\hat{Y}(f)=\hat{Y}(f+nf_e)\ \forall n$ by the same technique already applied to periodic signals in the time domain:

$$\text{TF}^{-1}\left[\hat{Y}(f)\right]=\int_{-\infty}^{\infty}\hat{Y}(f)\exp(j2\pi ft)df=\sum_{n=-\infty}^{\infty}\int_{-f_e/2}^{f_e/2}\hat{Y}(f)\exp[j2\pi t(f+nf_e)]df$$

$$=\left[\int_{-f_e/2}^{f_e/2}\hat{Y}(f)\exp(j2\pi ft)df\right]\sum_{n=-\infty}^{\infty}\exp[j2\pi tnf_e]$$

$$=\left[T_e\int_{-f_e/2}^{f_e/2}\hat{Y}(f)\exp(j2\pi ft)df\right]\sum_{k=-\infty}^{\infty}\delta[t-kT_e],$$

where the integral in brackets only has to be evaluated for the sole moments $t_k = kT_e = \dfrac{k}{f_e}$ where the samples are collected because the time Dirac comb only contains impulses at times t_k.

Therefore, we can put the result of the integral calculated at $t_k = k\,T_e$ inside the summation because it then becomes dependent of the index k when t is replaced by t_k:

$$\hat{y}(t) = \mathrm{TF}^{-1}\left[\hat{Y}(f)\right] = \sum_{k=-\infty}^{\infty} y_k\,\delta(t - kT_e),$$

which is a series of samples of value y_k with $y_k = \dfrac{1}{f_e}\displaystyle\int_{-f_e/2}^{f_e/2} \hat{Y}(f)\exp\left(j2\pi k\frac{f}{f_e}\right)df$.

Conclusion: the time signal whose FT is periodic in the frequency domain is a sampled signal (or discrete) in the time domain.

1.1.2. *Sampling (or Shannon's) theorem, anti-aliasing filtering and restitution of the continuous-time signal using the Shannon interpolation formula*

If the spectrum $Y(f)$ of the continuous time signal $y(t)$ has a bounded support smaller than $[-f_e/2, +f_e/2]$, namely equal to 0 outside of an interval narrower than $[-f_e/2, +f_e/2]$, $\hat{Y}(f) = f_e \sum_{n=-\infty}^{+\infty} Y(f - nf_e)$ can be replaced by $f_e\,Y(f)$ between the bounds of the integration interval $[-f_e/2, +f_e/2]$ in the previous computation of y_k with an upper bound $f_{max} < f_e/2$:

$$y_k = \frac{1}{f_e}\int_{-f_e/2}^{f_e/2} \hat{Y}(f)\exp\left(j2\pi f kT_e\right)df = \int_{-f_{\max}}^{f_{\max}} Y(f)\exp\left(j2\pi f kT_e\right)df = y(kT_e).$$

Nonetheless, in the case of a bounded spectrum in the interval $[-f_{\max}, +f_{\max}]$, itself included in the interval $[-f_e/2, +f_e/2]$, (Figure 1.8),

$\int_{-f_{\max}}^{f_{\max}} Y(f)\exp(j2\pi f kT_e)df$ can be replaced by $\int_{-\infty}^{\infty} Y(f)\exp(j2\pi f kT_e)df$, the inverse FT of $Y(f)$ evaluated at times kT_e or still the amplitude y_k of the samples. We can deduce the sampling (or Shannon's) theorem:

If a signal $y(t)$, whose spectrum $Y(f)$ is 0 outside $[-f_{max}\ f_{max}]$, is sampled with a sampling rate f_e twice larger than the bound f_{max}, the samples calculated by the inverse transform of the spectrum $\hat{Y}(f)$ of the sampled signal $\hat{y}(t)$ coincide with the values of the continuous-time signal $y(t)$ evaluated at sampling times $t_k = kT_e$.

In the frequency domain, this leads us to state that:

The spectrum of the original signal is preserved after sampling in the interval $[-f_e/2, +f_e/2]$ provided that it is bounded by a frequency f_{max} smaller than $f_e/2$, corresponding to the condition: $\boldsymbol{f_e > 2 f_{max}}$.

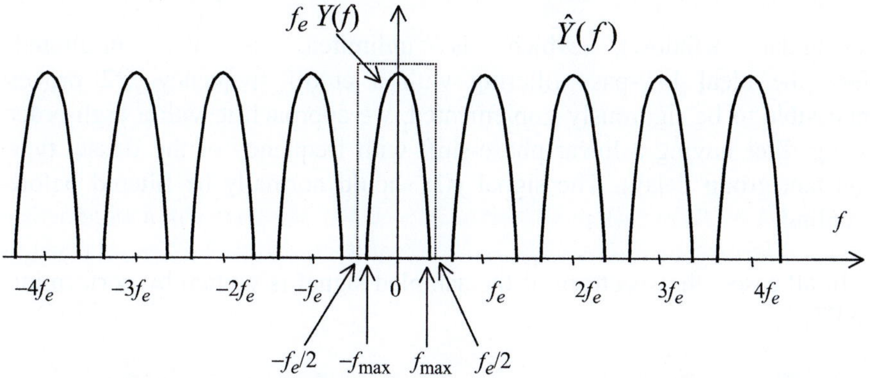

Figure 1.8. *Spectrum of a sampled signal which follows the Shannon theorem*

Conversely, if Shannon's theorem is not followed, the spectrum $\hat{Y}(f) = f_e \sum_{n=-\infty}^{+\infty} Y(f - nf_e)$, which is a sum of all the spectra of the continuous-time signal shifted by nf_e, exhibits in the interval $[-f_e/2, +f_e/2]$ at least part of the spectrum $Y(f \pm f_e)$ usually referred to as the aliased part (Figure 1.9):

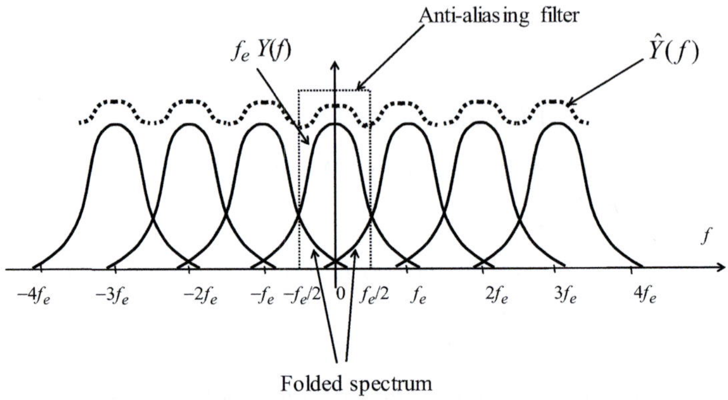

Figure 1.9. *Spectrum of a sampled signal for which the sampling theorem has not been followed*

An anti-aliasing low-pass filter is then necessary. In practice, it still is, because a real signal is limited in time and can therefore be considered as being the product of a rectangular window from $-T_0$ to $+T_0$ by an unlimited signal. Its spectrum then being the convolution with $\frac{\sin(2\pi T_0 f)}{\pi f}$ (FT of the rectangular window), which is unlimited, is also unlimited. Since the ideal low-pass filtering with a cutoff frequency $f_e/2$ proves impossible to be rigorously implemented, we approach it with a high-order analog filter having a linear phase-shift with frequency of the Bessel type (constant group delay). The signal $y(t)$ should normally be filtered before sampling.

In all cases, the spectrum of the sampled signal is written by performing the FT:

$$\hat{Y}(f)=\int_{-\infty}^{\infty} y(t)\sum_{k=-\infty}^{\infty}\delta(t-kT_e)\exp(-j2\pi f t)\,dt=\sum_{k=-\infty}^{\infty} y(kT_e)\exp(-j2\pi f kT_e).$$

If, and only if, the sampling theorem is applicable, spectra $\hat{Y}(f)$ and $f_e\,Y(f)$ are identical in the interval $\left[-\frac{f_e}{2},\frac{f_e}{2}\right]$. Therefore, $y(t)$ can be found from the inverse FT of $\hat{Y}(f)/f_e = T_e\hat{Y}(f)$ limited to the interval $\left[-\frac{f_e}{2},\frac{f_e}{2}\right]$, which is tantamount to implementing an ideal time-continuous low-pass filtering of the signal sampled at the cutoff frequency $\frac{f_e}{2}$, of transmittance zero outside $\left[-\frac{f_e}{2}, \frac{f_e}{2}\right]$:

$$y(t) = \int_{-f_e/2}^{f_e/2} T_e \sum_{k=-\infty}^{\infty} y(kT_e) \exp(-j2\pi f kT_e) \exp(j2\pi f t) df$$

$$= \sum_{k=-\infty}^{\infty} T_e \, y(kT_e) \int_{-f_e/2}^{f_e/2} \exp[j2\pi f(t - kT_e)] df,$$

which finally delivers the Shannon interpolation formula:

$$\boxed{y(t) = \sum_{k=-\infty}^{\infty} y(kT_e) \frac{\sin[\pi f_e(t - kT_e)]}{\pi f_e(t - kT_e)}.}$$

Remarkable conclusion: it is theoretically possible to recover the time-continuous signal $y(t)$ at any time t, merely from its samples $y(kT_e)$, if the sampling theorem can be applied; this requires knowledge of all the samples prior and after time t, considered with a decreasing weight when $|t - kT_e|$ increases.

This is achievable by means of numerical computation (samples digitized and stored; see section 1.4.7) but with a finite number of samples in practice; it further enables the interpolation to be achieved by oversampling ($2^m - 1$ additional samples calculated between each original samples, rejecting the sampling frequency at $2^m f_e$).

1.1.3. *Discrete Fourier series (or transform); "fast Fourier transform" (FFT) and discrete cosine transforms (DCT)*

It is assumed that the signal $y(t)$ is sampled N times over a period $T_0 = 1/f_0$, which is considered to be the period of the signal. The resulting signal is periodic and sampled with sampling period $T_e = \frac{T_0}{N}$ (Figure 1.10):

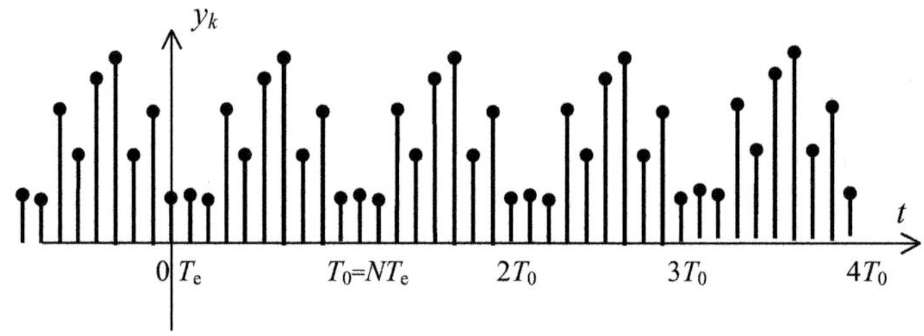

Figure 1.10. *Sampled periodic signal*

The spectrum is thus discrete and periodic of period $f_e = N f_0$ in the frequency domain. In order to obtain the "discrete Fourier transform", we calculate the coefficients Y_n of the Fourier series decomposition of $T_e\, y(t)$ which has been sampled or, in other words, multiplied by a Dirac comb $\sum_{k=-\infty}^{\infty} \delta(t-kT_e)$, over a time period $T_0 = NT_e$:

$$Y_n = \frac{1}{T_0}\frac{T_0}{N}\int_0^{T_0}\sum_{k=0}^{N-1}\delta(t-kT_e)\,y(t)\exp\left(-j\frac{2\pi nt}{T_0}\right)dt = \frac{1}{N}\sum_{k=0}^{N-1} y(kT_e)\exp\left(-j\frac{2\pi kn}{N}\right),$$

which represents the complex amplitude of the spectrum line of $y(t)$ sampled at frequency $nf_0 = \frac{n}{T_0} = \frac{n}{NT_e} = \frac{n}{N} f_e$ and periodic of frequency period f_e.

The spectrum can thus be written as $\sum_{n=0}^{N-1} Y_n\,\delta(f - \frac{n}{N} f_e)$ when limited to an interval [0 f_e], (Figure 1.11).

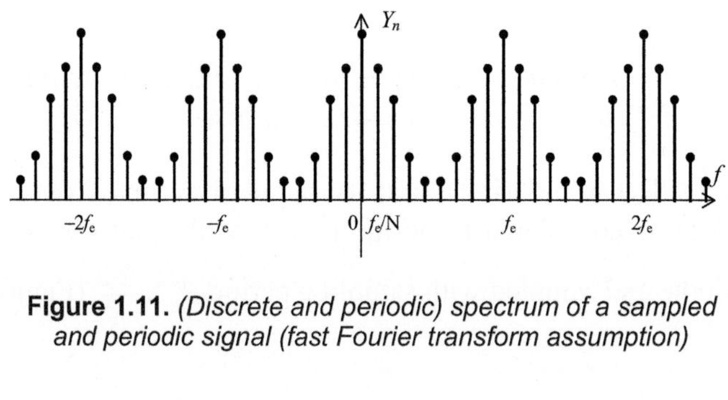

Figure 1.11. *(Discrete and periodic) spectrum of a sampled and periodic signal (fast Fourier transform assumption)*

We can even restrict ourselves to $N/2$ lines because given that $y(t)$ is real, $|Y_n|$ is even and $\mathrm{Arg}\{Y_n\}$ is odd; Y_n can thus be deduced in all other frequency half-periods, especially from $-f_e/2$ to 0 and from $f_e/2$ to f_e.

We can calculate the inverse discrete transform of spectrum Y_n limited to the frequency period f_e, since the spectrum is periodic, in order to find the time samples:

$$\int_0^{f_e} \sum_{n=0}^{N-1} Y_n \, \delta(f - \frac{n}{N} f_e) \exp(j2\pi k T_e f) = \sum_{n=0}^{N-1} Y_n \exp\left(j2\pi \frac{kn}{N} \right).$$

Hence, finally, the pair of discrete Fourier series can be written as follows by establishing that $y[k] = y(kT_e)$ and $Y[n] = Y(nf_e) = Y_n$:

$$Y[n] = \frac{1}{N} \sum_{k=0}^{N-1} y[k] \exp\left(-j2\pi \frac{kn}{N} \right)$$

$$y[k] = \sum_{n=0}^{N-1} Y[n] \exp\left(j2\pi \frac{kn}{N} \right)$$

These expressions can be computed by an algorithm that makes use of the symmetry properties of exponentials (*fast Fourier transform, abbreviated by FFT*) and are very often available in digital oscilloscopes. A time window is normally required to limit the discontinuity between the first and the last time samples, which exists due to the quasi-systematic lack of true periodicity. In effect, without a window, this discontinuity introduces a modification in the spectrum. The most used windows are the Hanning $= \frac{1}{2} - \frac{1}{2} \cos\left(\frac{2\pi n}{N-1} \right)$, the Hamming $= 0.54 - 0.46 \cos\left(\frac{2\pi n}{N-1} \right)$ windows, those with a flat-top from $n = N/4$ to $3N/4$ showing a decrease on both sides of the peak, the Blackmann $= 0.42 - 0.5 \cos\left(\frac{2\pi n}{N-1} \right) - 0.08 \cos\left(\frac{4\pi n}{N-1} \right)$ or still the triangular windows; moreover, the rectangular window does not bring any change (windowing is covered in more detail in sections 1.4.4.1 and 1.4.6.5).

There are other existing spectrum calculations, in which complex exponentials are replaced by cosines (from DCT-I to DCT-IV). They make it possible to better "focus" the spectrum in low frequencies; they still take advantage of computational methods using fast algorithms and lend themselves better to data compression by way of truncating the spectrum in higher frequencies. In addition, one way of solving the discontinuity problem between samples of ranks 0 and N-1, much less damaging to the integrity of the signal, is based on the discrete MDCT (Modified Discrete Cosine Transform) which employs twice the number of samples:

$$Y[n]=\sum_{k=0}^{2N-1} y[k]\cos\left[\frac{\pi}{N}\left(k+\frac{1}{2}+\frac{N}{2}\right)\left(n+\frac{1}{2}\right)\right] \quad \text{for } n \in [0,\ N-1].$$

The inverse IMDCT transformation (Inverse Modified Discrete Cosine Transform) allows us to find $2N$ samples, using the expression:

$$y[k]=\frac{1}{N}\sum_{n=0}^{N-1} Y[n]\cos\left[\frac{\pi}{N}\left(k+\frac{1}{2}+\frac{N}{2}\right)\left(n+\frac{1}{2}\right)\right] \quad \text{for } k \in [0,\ 2N-1].$$

This type of method provides a means to manage continuously streaming data, such as audio streams, or those with redundant samples, and also to store them in their spectral form after compression or to reproduce the compressed data as a function of the time index (treatments used in MPEG protocols, an abbreviation of "Moving Picture Experts Group", like MPEG-II, more commonly known as "mp3").

1.2. Discrete time–continuous time interface circuits

Since one of the terminations of these circuits receives or generates continuous-time signals, the analytical methods are here again the FT and the Laplace transform.

1.2.1. *Real sampler*

A real sampler has a non-zero sampling time αT_e, equal to a fraction α of the sampling period $T_e = 1/f_e$ ($0< \alpha <1$) and can be represented by the following schematics (Figure 1.12):

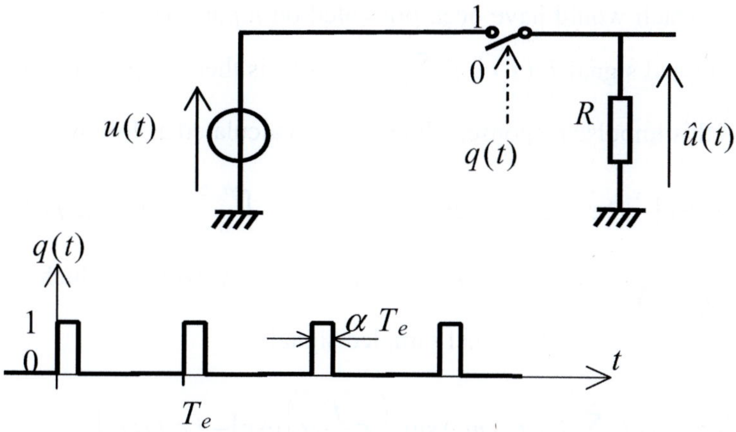

Figure 1.12. *Simple sampler and its control signal*

The clock signal $q(t)$ controls the real sampler. Its spectrum $Q(f)$ is shown in Figure 1.13, together with that of the input signal, assumed to verify the sampling theorem condition.

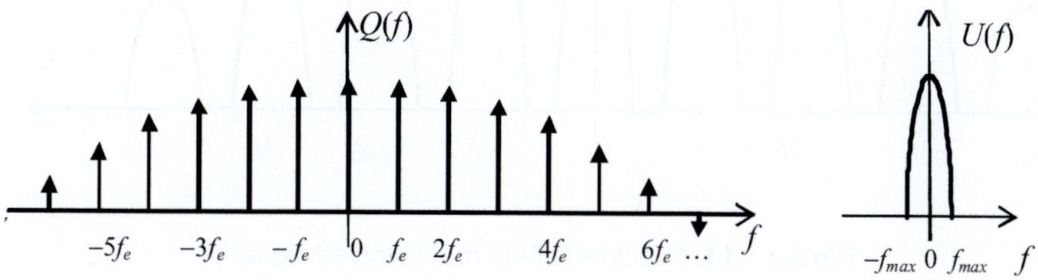

Figure 1.13. *Spectra of the control signal and of the signal before the sampler, assumed to comply with Shannon's theorem condition $f_{max} < f_e/2$*

It should be noted that the actual sampler shows an impulse response which is none other than the only rectangular pulse of $q(t)$ located between 0 and T_e and that the series of impulses observed in $q(t)$ is nothing other than the response of the actual system to a Dirac comb characteristic of ideal sampling, which would have been presented on input. The spectrum of the ideally sampled signal $\hat{U}(f) = f_e \sum_{n=-\infty}^{+\infty} U(f - nf_e)$ is then simply multiplied by the FT of this impulse response, which can be calculated as follows:

$$\int_0^{\alpha T_e} f_e \exp[-j2\pi\, f t]dt = f_e \exp[-j\pi\, f \alpha T_e] \int_{-\alpha T_e/2}^{\alpha T_e/2} \exp[-j2\pi\, f t']dt'$$

$= \alpha \dfrac{\sin(\pi f \alpha T_e)}{\pi f \alpha T_e} \exp[-j\pi\, f \alpha T_e] = \alpha \operatorname{sinc}(\pi f \alpha T_e) \exp[-j\pi\, f \alpha T_e]$, which finally gives for the spectrum of the real sampled signal:

$$\hat{U}(f) = \alpha f_e \sum_{n=-\infty}^{+\infty} U(f - nf_e) \operatorname{sinc}\left(\pi \frac{f}{f_e} \alpha\right) \exp[-j\pi\, f \alpha T_e].$$

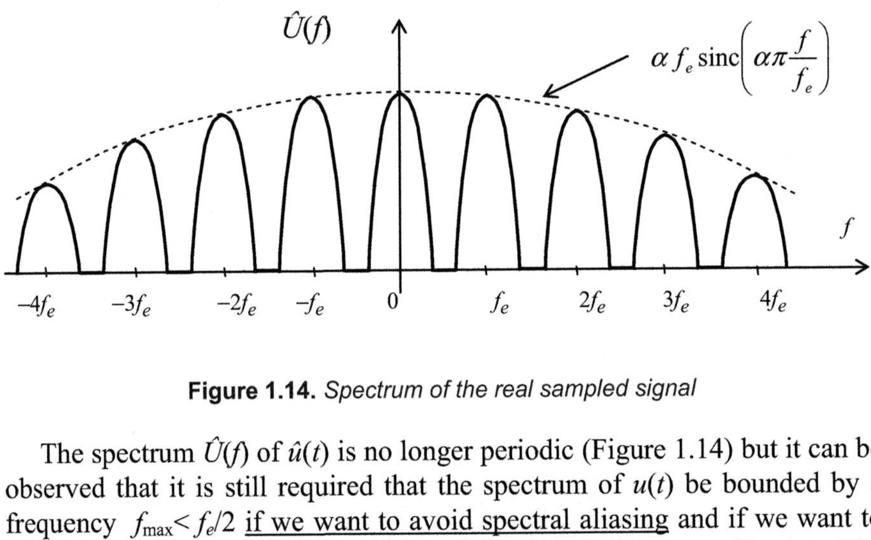

Figure 1.14. *Spectrum of the real sampled signal*

The spectrum $\hat{U}(f)$ of $\hat{u}(t)$ is no longer periodic (Figure 1.14) but it can be observed that it is still required that the spectrum of $u(t)$ be bounded by a frequency $f_{max} < f_e/2$ if we want to avoid spectral aliasing and if we want to be able to recover the original signal using simple low-pass filtering. The sampling theorem therefore applies in the same way as for the ideally sampled signal.

The components of the spectrum located around the multiples of the sampling frequency f_e are attenuated by the factor $\operatorname{sinc}\left(\alpha\pi \frac{f}{f_e}\right)$ with respect to that located in the neighborhood of $f = 0$. The factor $\exp\left[-j\pi f \alpha T_e\right]$ corresponds to the introduction of a delay $\alpha T_e/2$, which is half the sampling duration.

1.2.2. *Sample-and-hold circuit*

In many cases, the sampler is replaced by a sample-and-hold circuit, either deliberately in order to leave the data during the entire period T_e at the input of a digital system or because the system works in a sequential fashion and stores data during one clock period, as is often the case in a digital-to-analog converter (DAC) just before the conversion step. It is thus a function capable of operating both on input and on output of a digital system, functioning as an interface between discrete-time and continuous-time signals. It can be described by a sampler in which the resistance R of the previous schematics has been replaced by a capacitance C, which acts as analog memory (Figure 1.15). It is charged at the new sampled value of the input signal at each period T_e and maintains this charge during the entire period T_e. The sampled-and-held signal is thus deduced from the previous one by giving α the value of 1, which allows its spectrum to be immediately obtained by replacing α by 1 in the expression of $\hat{U}(f)$ of the preceding section.

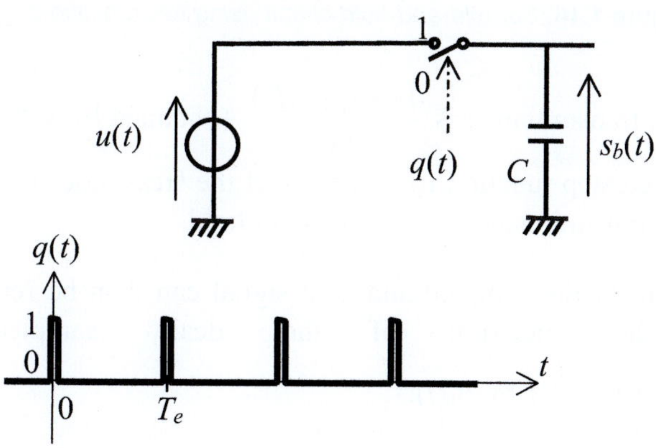

Figure 1.15. *Sample-and-hold circuitry and its control signal*

The practical implementation of sample-and-hold circuitry requires a voltage source before capacitance C and a follower of infinite input impedance so that the capacitor does not discharge when the switch is open.

For example, with the circuit in Figure 1.16. If the switching times enforced are too short for operational amplifiers to be able to respond fast enough, a switch making use of four diodes (see Chapter 1 of Volume 1, [MUR 17a]) is employed.

The transmittance of the sample-and-hold circuit (or zero-order hold) is then simply the FT of its impulse response, which is a rectangular pulse of amplitude 1 in the semi-open interval $[0, T_e[$:

$$B_0(f) = \int_0^{T_e} \exp[-j2\pi f t]\, dt = \frac{\sin(\pi f T_e)}{\pi f} \exp[-j\pi f T_e] = T_e \operatorname{sinc}(\pi f T_e) \exp[-j\pi f T_e].$$

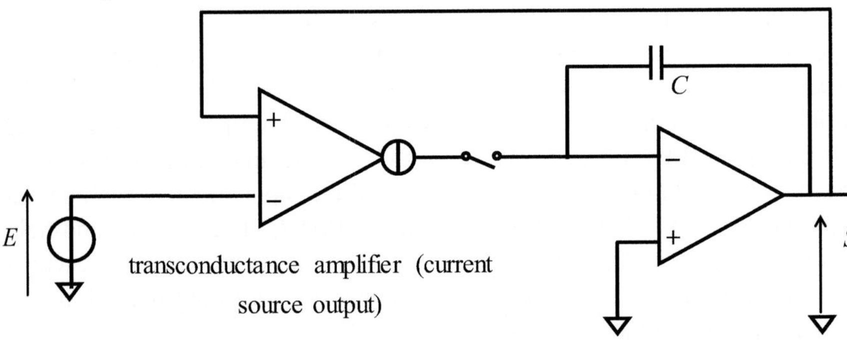

Figure 1.16. *Sample-and-hold circuit using two amplifiers*

The Laplace transmittance is $\frac{1-\exp(-sT_e)}{s}$ (obtained by subtracting the LT of the delayed step function from the LT of the Heaviside step function), often called the transmittance of the zero-order hold.

The spectrum of the sampled-and-held signal can then be recovered by multiplying the spectrum of the ideally sampled signal $\hat{U}(f) = f_e \sum_{n=-\infty}^{+\infty} U(f - nf_e)$ by $B_0(f)$:

$$S_b(f) = \operatorname{sinc}\left(\pi \frac{f}{f_e}\right) \exp\left[-j\pi \frac{f}{f_e}\right] \sum_{n=-\infty}^{+\infty} U(f - nf_e).$$

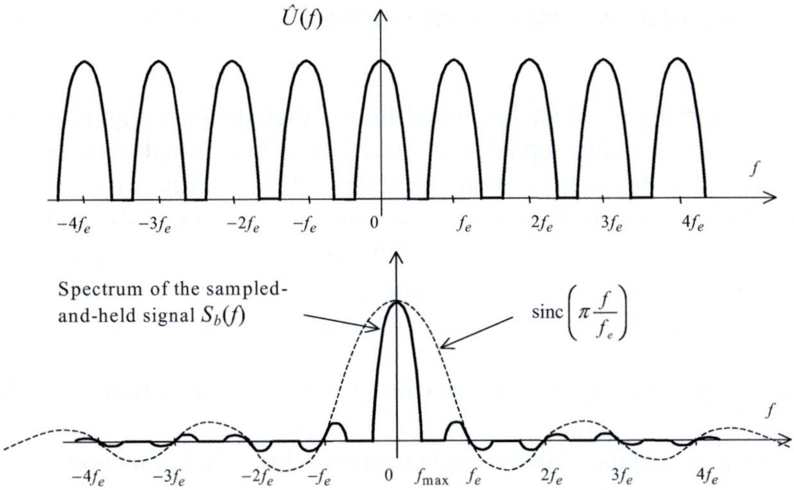

Figure 1.17. *Spectra of the ideally sampled signal (on top) and of the sampled-and-held signal (at the bottom)*

The components of the spectrum located around the multiples of the sampling frequency nf_e (or satellites) are now attenuated in modulus by the factor $\operatorname{sinc}(\alpha\pi f/f_e)$, which is 0 at any sampling frequency nf_e except at $f = 0$ (Figure 1.17).

The recovery of the analog signal can therefore be carried out more easily because the satellites of the original spectrum located around the multiples of the sampling frequency are already attenuated compared to those of the spectrum of the ideally sampled signal.

It is, nevertheless, preferable to correct the attenuation in $\operatorname{sinc}\left(\pi\frac{f}{f_e}\right)$ in the bandwidth using inverse transmittance filtering, that is in $\dfrac{1}{\operatorname{sinc}\left(\pi\frac{f}{f_e}\right)}$ for $|f| < f_{\max}$ then with a strong attenuation for $|f| > f_{\max}$. The first operation can be performed with a rational approximation of the transfer function based on specialized circuits, before carrying out the low-pass filtering that will eliminate the satellites of the spectra at $|f| > f_{\max}$. This problem is discussed more generally in the following section.

1.2.3. *Interpolation circuits and smoothing methods for sampled signals*

The recovering of an analog signal from a digital signal requires a DAC which delivers a value updated at every sampling or conversion clock period. If this analog value is maintained throughout the whole of the period, the transfer function is that of a zero-order hold, as previously studied. In order to improve the smoothness of these data, it proves beneficial to implement low-pass filtering and filtering for correcting the response of the zero-order sample-and-hold circuit using its inverse transmittance $\left[\operatorname{sinc}\left(\pi\frac{f}{f_e}\right)\right]^{-1}$ in the bandwidth (for $|f| < f_{\max}$). One solution involves interposing a digital filter implementing this dual function before conversion.

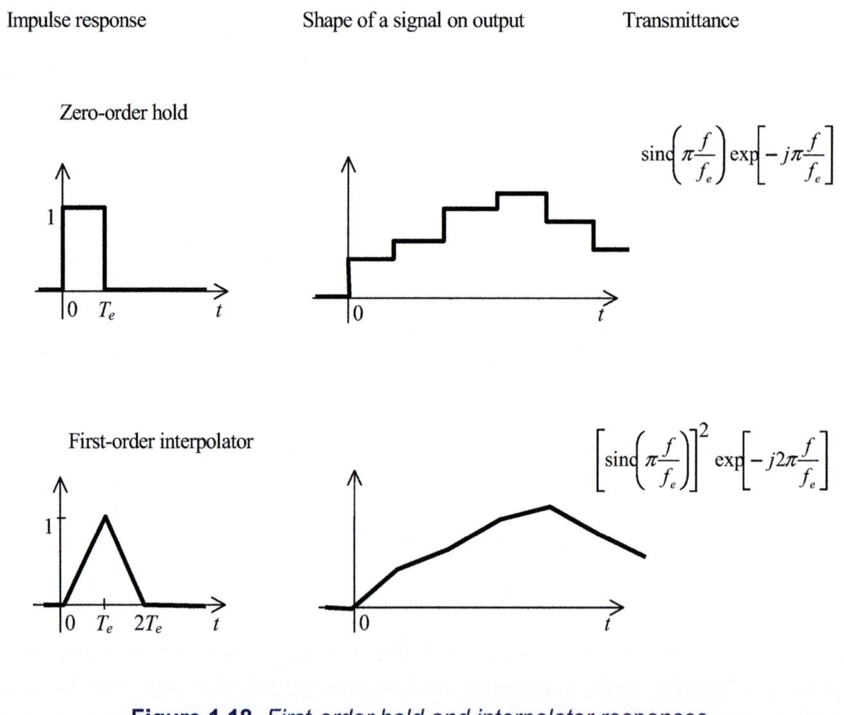

Figure 1.18. *First-order hold and interpolator responses*

Another solution involves using a kth-order interpolator sampler of transmittance $\left[\operatorname{sinc}\left(\pi\frac{f}{f_e}\right)\right]^{k+1}$. In the time domain, this filtering achieves interpolation-based smoothing of the output signal of the sample-and-hold circuit, which is a step function. The corresponding impulse responses are inferred from that of the zero-order hold using time convolution, because raising a transfer function to the $k+1$ power in the frequency domain corresponds to the k successive convolutions of the initial impulse response $h_0(t)$ in the time domain: $h_1(t)=\frac{1}{T_e}\int_{-\infty}^{+\infty} h_0(\tau)h_0(t-\tau)d\tau$; $h_2(t)=\frac{1}{T_e}\int_{-\infty}^{+\infty} h_0(\tau)h_1(t-\tau)d\tau$; $h_3(t)=\frac{1}{T_e}\int_{-\infty}^{+\infty} h_1(\tau)h_1(t-\tau)d\tau$, etc. From $h_0(t)$, a rectangular pulse equal to 1 between 0 and T_e, these impulse responses then become a triangle (Figure 1.18), arcs of parabolas, arcs of third, forth-degree polynomial functions and so on. The number of segments is $k+1$, and the impulse response has a total duration $(k+1)T_e$.

For interpolators of orders higher than 1, the operation cannot be carried out in a rigorous way using purely analog techniques. It is also misleading to dispose sample-and-hold circuits in cascade similar to the circuit in Figure 1.16 because the sampling is renewed at each stage. This may be interpreted as the consequence of the stationary approximation that is made to calculate the transfer function of the sample-and-hold circuit. However, in reality, this system is not stationary because the duration of the rectangular pulse observed on output is equal to the sampling period only in the case where the impulse sent on input coincides with a sampling time; it is smaller otherwise. A solution consists of generating the continuous-time impulse response, to assign to each portion of duration T_e an amplitude corresponding to the sample being considered $u[n]$, $u[n-1]$ and so on by way of a DAC multiplier (see Chapter 2 in this Volume), and to perform the sum of individual contributions, as in the example hereafter (Figure 1.19).

For a second-order interpolator, it is necessary to generate three shifted arcs of parabola and to place them on the input of three DAC multipliers, and then to carry out the sum. This technique is particularly well suited to the output of DACs.

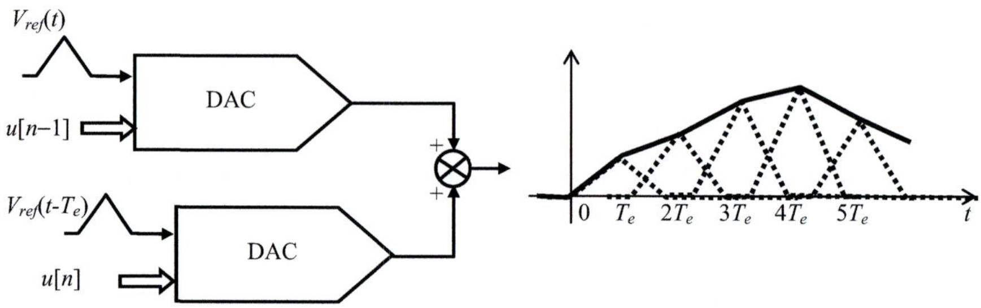

Figure 1.19. *First-order interpolator using two DAC multipliers*

An alternative solution implemented by a numerical method detailed further in the text involves increasing the sampling rate by making use of Shannon's interpolation formula to calculate intermediate samples, at a frequency N times superior to f_e.

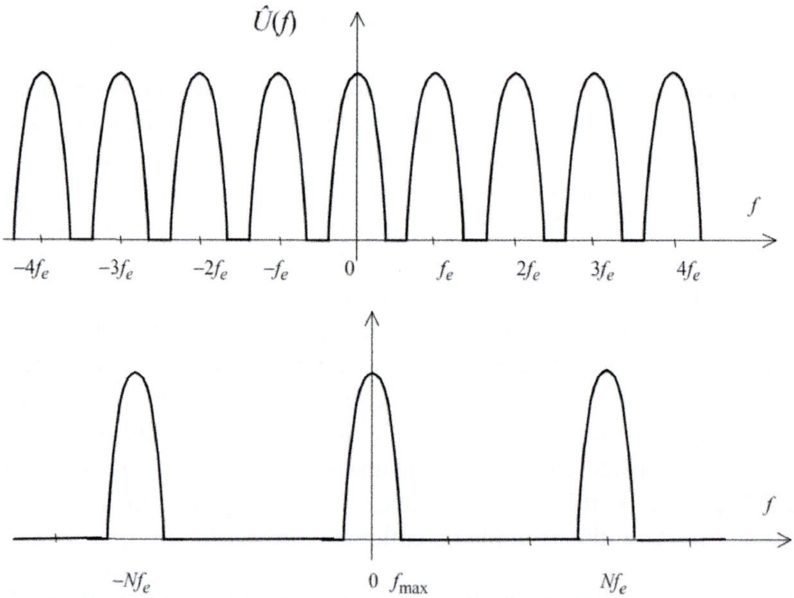

Figure 1.20. *Spectra of either sampled (top) and over-sampled (bottom) signals using oversampling or of sampled (bottom) and sub sampled (top) signals using decimation*

This solution is a means to shift satellites from the spectrum $\hat{U}(f)$ of the sampled signal away from one another by replacing $U(f - nf_e)$ by $U(f - nNf_e)$ (Figure 1.20). It facilitates the low-pass filtering needed to recover the original signal because Nf_e is much larger than f_e and the order of the filter can be reduced in significant proportions. In some cases, such as on output of DACs transforming digital signals originating from a compact disc (CD), the filter even becomes implementable whereas it was not in the absence of oversampling; N being generally equal to 16, 32, 64 or even 256, which amounts to calculating 15, 31, 63 or 255 additional samples between each original sample.

Last, it can be considered that the non-stationarity of the sample-and-hold circuitry becomes negligible if signals vary quite slowly compared to the sampling period, which can be regarded as another interpretation of the sampling theorem when it is verified. In this case, sampled filtering which involves repeating the sampling operation N times, that is to say performing a rolling average, becomes even more interesting due to a very easy implementation based on a digital filter if the signal is digitized immediately after the sample-and-hold circuit. The simplest filtering is performed by a comb filter, so called because of its impulse response presented in Figure 1.21. Other more powerful low-pass filtering (section 1.4.6.5) is possible and may be preferable.

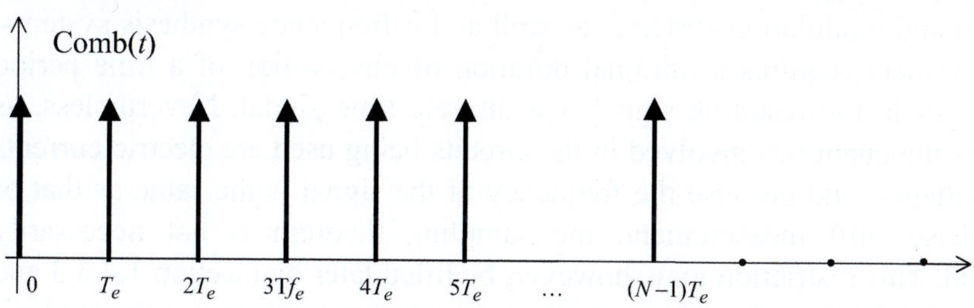

Figure 1.21. *Impulse response of the comb filter, comprising N successive samples of the same magnitude, followed by zero-amplitude samples.*

The previous solution is useful when oversampling has occurred beforehand, and when we must then operate a downsampling to restore the signal, which implies that the sampling theorem remains verified. This is the case in sigma-delta modulators where the condition $f_e > 2\ f_{\max}$ is largely surpassed as a consequence of the frequency utilized to clock the functioning of the 1-bit DAC (see Chapter 2), of a factor N much higher than 100. The digital system that processes the flow of 1-bit encoded data on output of the sigma-delta modulator implements both low-pass filtering which is necessary to attenuate the conversion noise as well as the transcoding of 1-bit-encoded data toward n-bit encoding. These techniques are discussed in more detail in section 1.4.7 and in Chapter 2. They allow us to improve the basic "decimation" operation, which simply consists of taking a single sample out of N. Since such an operation is accompanied by a decrease in the sampling frequency with the same ratio N, this represents a subsampling which involves a possible overlapping of the original spectrum with the satellites depicted at the top of Figure 1.20 in the frequency domain if those depicted at the bottom of this same figure are too close to the portion of the spectrum centered on the zero frequency. As a conclusion, the validity of Shannon's theorem has to be verified every time the sampling frequency is changed, especially when it is downscaled.

1.3. Phase-shift measurements; phase and frequency control; frequency synthesis

Sample-and-hold circuits are fundamental for measuring the phase shift between two signals, and they provide a basis for phase shift and frequency control and modulation systems, as well as for frequency synthesis systems. Measurement requires a minimal duration of observation of a time period and as such, the resulting signal is a discrete time signal. Nevertheless, as long as the quantities involved in the circuits being used are electric currents and voltages, and because the frequency of the signal is the same as that of the phase shift measurement, the sampling theorem is not necessarily verified. This restriction may, however, be lifted later on (section 1.4.3.3 and section 1.4.5). As a first step, an adequate solution to carry out the analysis then involves using a continuous-time approximation of the response of the sample-and-hold circuit.

1.3.1. *Three-state circuit for measuring the phase shift*

The basic circuit should allow us to obtain a linear relation between the voltage or the output current and the phase shift existing between two signals, and this should hold even when the phase shift changes sign. This requires a three-state switch instead of two-state for sample-and-hold circuits previously studied and two voltage and current sources. Other simpler systems, with two states only, provide a linear relation over a more limited phase shift range and are incapable of distinguishing lead from lag phase shift. These are functions based on an exclusive OR circuit that will not be considered here.

First, let $\varphi(kT_e)$ be the algebraic phase difference for any period kT_e (k integer) between two sinusoidal signals $s_2(t)$ and $s_1(t)$ of the same period T_e. This phase difference is defined by $2\pi\frac{\Delta t}{T_e}$, where Δt represents the time interval between two crossings of zero or predetermined amplitude, in the same direction. Without loss of generality, we can also consider signals $s_1(t)$ and $s_2(t)$ to be rectangular signals and consider Δt as the interval between the two rising edges because the addition of a comparator to both channels is sufficient to obtain such signals from sinusoidal signals, or even for more complex alternative signals; at least in the case in which the passage through a particular amplitude in one direction is achieved only once per period. In this case also, the phase shift $\varphi(kT_e)$ between $s_2(t)$ and $s_1(t)$ is defined for each period kT_e. It is positive if $s_2(t)$ is leading $s_1(t)$ and negative if $s_2(t)$ is lagging $s_1(t)$ as shown in Figure 1.22.

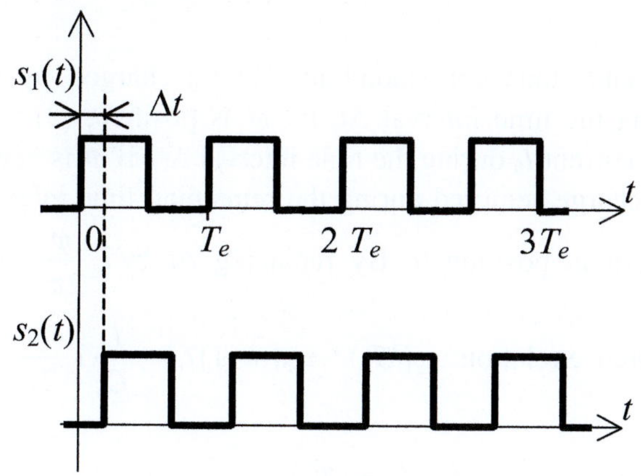

Figure 1.22. *Square signals shifted by Δt*

We will consider only the three-state system to be capable of taking into account the sign of the phase shift (circuit in Figure 1.23) and showing a linear response range $\pm 2\pi$ [CD 4046]. In this circuit, the switches K_1 and K_2 are controlled by the time step Δt: K_2 or K_1 is in position 1 (or H) during Δt depending on whether the phase φ is respectively positive (phase lead) or negative (phase lag); K_1 and K_2 are in zero position (or L) outside the interval of duration Δt between edges of the same direction.

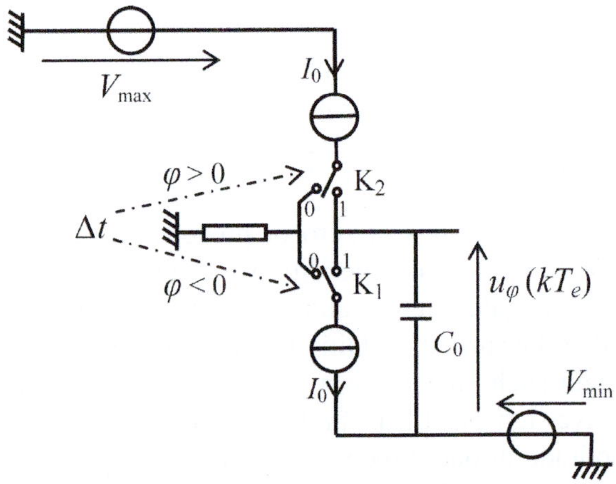

Figure 1.23. *Circuit for measuring the integral of the phase shift between square signals, controlled by the binary signals K_1 and K_2, either one being 1 during Δt according to the sign of φ*

The three states thus correspond to: (1) C_0 charged by the constant current I_0 during the time interval Δt, if φ is positive; (2) C_0 discharged by the constant current I_0 during the time interval Δt, if φ is negative; (3) C_0 preserving the charge acquired during the remaining time interval when the two switches are in position 0. By replacing Δt by $T_e \dfrac{\varphi}{2\pi}$, we have the following recurrence relation: $u_\varphi(kT_e) = u_\varphi[(k-1)T_e] + \dfrac{I_0}{C_0} T_e \dfrac{\varphi}{2\pi}$.

The increase in voltage $\dfrac{I_0}{C_0} T_e \dfrac{\varphi}{2\pi}$ results from the charging (or discharging) of the capacitor during a fraction Δt of the period, figured by the solid line in Figure 1.24:

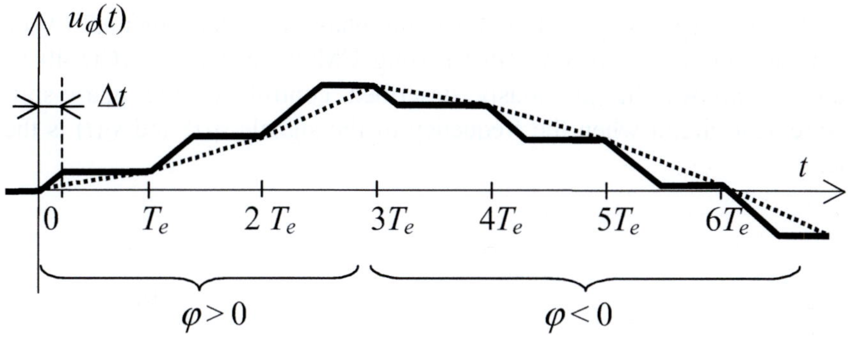

Figure 1.24. *Signal output of the circuit for measuring the integral phase shift (solid line) and its continuous-time approximation (dotted line)*

This increase is indeed proportional to the phase shift φ, regardless of its sign. Since the sample period T_e merges with the period of electric signals $s_1(t)$ and $s_2(t)$, the sampling theorem is not verified. The most fruitful approximation then involves replacing the actual variation shown with a solid line in Figure 1.24 by the dashed-line segments that join the values of $u_\varphi(kT_e)$ at each time kT_e, which amounts to replacing $u_\varphi(kT_e)$ by $u_\varphi(t)=\dfrac{I_0}{2\pi C_0}\int \varphi(t)\,dt$, which is obtained by integrating $\dfrac{I_0}{C_0}\dfrac{\varphi(t)}{2\pi}$ that is to say the slope of each segment. This is all the more justified that the capacitance plays the role of integrator, and that there is no zero resetting of its charge at each period. It thus keeps the memory of the previous phase shifts, which corresponds to integration in the continuous-time domain. In the event it is desirable to change the sign of the relation between the phase comparator input and output signals, it suffices to change the connection of the capacitance C_0 at the polarization source $V_{\min}$ and to replace it with a connection to the polarization source $V_{\max}$.

Since this approximation consists of replacing the discrete time variable kT_e by the continuous-time variable t, it also makes it possible to assimilate $\varphi(t)$ with the instantaneous phase of a sinusoidal signal $s(t) = \sin[\varphi(t)]$, for example, or with the modulated phase shift $\varphi_m(t)$ of a fixed-frequency signal $\sin[2\pi\, f_0\, t + \varphi_m(t)]$, and therefore to consider all quantities as if they were dependent on the continuous time t.

The control of switches K_1 and K_2 is achieved based on a sequential logical system making use of synchronous positive-edge triggered D flip-flops (Figure 1.25). A typical circuit is the phase-shift detector no.2 of the 4046-based phase-locked loop (PLL) using CMOS technology [CD 4046] because it allows the previously described control for the phase-shift measurement circuit when the frequency of the signals $s_1(t)$ and $s_2(t)$ is the same.

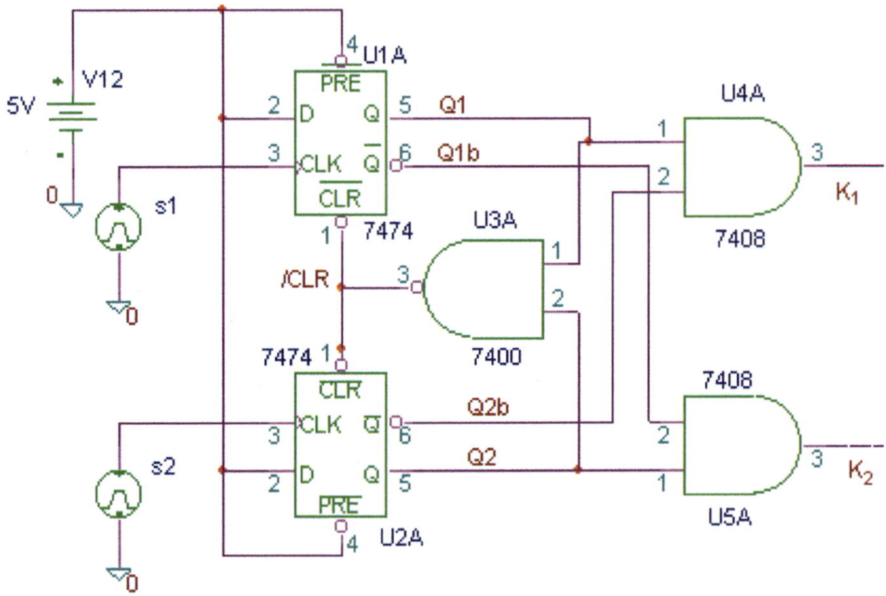

Figure 1.25. *Operation diagram of the control circuit of the phase-shift detector no.2 of the 4046 circuit*

In cases where the frequencies are identical, it is shown that the circuit of Figure 1.25 is capable of activating K_1 or K_2 to state H for the duration Δt according to the sign of the phase shift. For example, in the case of negative phase shift (s_2 lagging s_1) of Figure 1.26, Q1 first reaches H followed by Q2 after time Δt, while they were at state L previously as it will be shown due to the resetting. During this duration Δt, the output K_1 thus shifts to H, while K_2 remains L. Then the transition from Q2b to L brings K_1 back to L, while K_2 remains L, because of Q1b. This normal sequential operation of flip-flops ceases when the transition from Q2 to H is effected, because Q1 is also H at that moment, then the gate U3A causes the resetting of the flip-flops, making Q1, Q2, K_1 and K_2 shift to L (see Figure 1.26). The opposite applies if s_2 is leading s_1, which results in the circuits operating as full phase and frequency detectors (PFDs) as demonstrated further in the case of non identical input frequencies. The output signals of the U3A gate and of the flip-flop which receives the input signal lagging the other are impulses whose duration is determined by switching and propagation times of the circuits. Their duration can be increased by introducing "buffer" or inverter circuits (in even numbers) between the output of the NAND U3A gate and the reset (/CLR) input of the flip-flops.

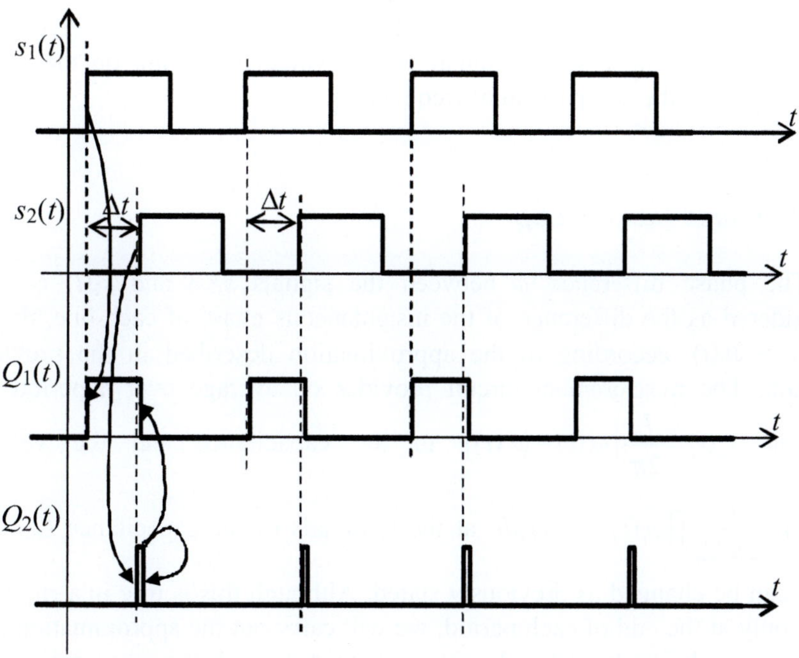

Figure 1.26. *Sequence of the input and output signals of the flip-flops in the case of signals with the same frequency and shifted by a lag* Δt

In addition, the circuit in Figure 1.25 operates as a frequency detector when the frequencies of $s_1(t)$ and $s_2(t)$ are different, respectively corresponding to periods T_1 and T_2. The phase shift can no longer be defined but it is always possible to reason based on the interval Δt between two rising edges of signals $s_1(t)$ and $s_2(t)$. Assuming that we start with synchronous signals and suddenly increase the frequency of $s_2(t)$ for instance, the first rising edge appearing on one of the inputs of the circuit of Figure 1.25 will be that of $s_2(t)$ followed on the other input by that of $s_1(t)$

with a delay $\Delta t = T_1 - T_2$, which will lead to reset (/CLR) the two flip-flop outputs Q1 and Q2 (Figure 1.25) and to stop the capacitance C_0 from charging. This process will be repeated with increasing delays $2(T_1 - T_2)$, $3(T_1 - T_2)$ and so on, until the delay becomes greater than T_2 and until the same process resumes starting with minimal delay. Since the charge of the capacitance C_0 increases for every duration Δt, $u_\varphi(t)$ will have had time to reach either the minimal saturation voltage $V_{\min}$, or the maximal saturation voltage $V_{\max}$. The effect is reversed in the case of a contrary condition on frequencies, which is reflected by the circuit operating as a frequency detector delivering a binary signal with a value depending on the status resulting from the comparison of frequencies.

1.3.2. *Phase-locked loop*

The phase difference φ between the signals $s_1(t)$ and $s_2(t)$ can be considered as the difference of the instantaneous phase of each one, that is $\varphi_1(t) - \varphi_2(t)$, according to the approximation described in the previous section. The measurement circuit provides on average over a period the current $i_\varphi(t)=\frac{I_0}{2\pi}[\varphi_1(t)-\varphi_2(t)]$ to the capacitance and the voltage $u_\varphi(t)=\frac{I_0}{2\pi C_0}\int[\varphi_1(t)-\varphi_2(t)]dt$ at the terminals of the capacitance C_0. The sign can be changed as previously stated. Although this is true in a rigorous way only at the end of each period, we will carry out the approximation that the measured signals $i_\varphi(t)$ and $u_\varphi(t)$ are indeed dependent of the continuous-time variable t. Considering the equivalent subtractor obtained after Laplace transform, we arrive at the following operation diagram (Figure 1.27), in which the transmittance $A_1(s)$ always contains a factor $\frac{I_0}{C_0}$, homogeneous to a voltage per time unit, because of the flow of current in the capacitance C_0 during a time interval that is at most the period T_e of the input signal. This case corresponds to a phase difference of 2π, and the corresponding variation in voltage $\Delta V_{2\pi}$ defined by $\frac{I_0 T_e}{C_0 T_e}=\frac{\Delta V_{2\pi}}{T_e}$ must be limited to a value smaller or at most equal to the difference of the polarization voltages of the circuits $V_{\max} - V_{\min}$ otherwise the risk of introducing non linearity could be incurred, reducing the usable phase-shift range. Most of the time, the value $\frac{I_0 T_e}{C_0 T_e} = \frac{V_{\max} - V_{\min}}{T_e}$ will be selected to allow for optimal sensitivity. However, as discussed further in the text, it proves useful in some cases to decrease the sensitivity, and to this end, we will be able to adopt a smaller $\Delta V_{2\pi}$ value by reducing the ratio I_0 / C_0.

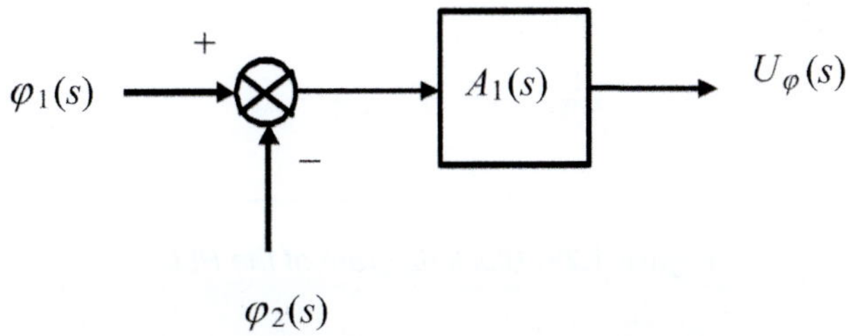

Figure 1.27. *Integrated phase subtractor*

The previous circuit diagram can be completed by a correction loop filter making use of a passive network and possibly one or two operational amplifiers, a *voltage-controlled oscillator* (VCO) of actual transmittance K_f followed possibly by a divide-by-N frequency divider if it is desirable to achieve a closed-loop multiply-by-N frequency multiplier (see Figure 1.28), and a functional block enabling the phase-shift frequency on the VCO output to be transformed by taking the Laplace transform of $\varphi(t) = 2\pi \int_0^t f(t')dt'$, equal to $\frac{2\pi}{s}$ $F(s)$ (with f and F being respectively the time dependent frequency and its LT). The control is then completed by unitary feedback.

In the absence of correction, the loop transmittance would be s^{-2}, with two poles on the imaginary axis, and would lead to an unstable closed-loop system. It is demonstrated that with a first-order phase-lead correction (see exercise 1.6.4), the closed-loop transmittance $H_\varphi(s)$ is of the second-order with a natural frequency ω_n and a damping factor ζ. This natural frequency has to be compared with the frequency variation of the phase or frequency imposed on the system input or with the modulation frequency. The parameters

ω_n and ζ must therefore be adapted to the necessities of the frequency response and of the desired time responses, whether the main function of the system is to operate as a modulator, a demodulator or simply for phase or frequency control. The general block diagram of the loop is given in Figure 1.28.

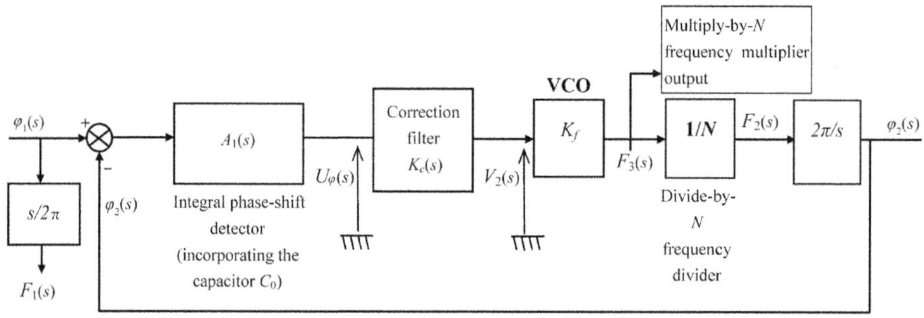

Figure 1.28. *Block diagram of the PLL*

The VCO can either be a sinusoidal oscillator in which the capacitance controlling the oscillation frequency is partly determined by a "Varicap" diode (see Chapter 1 of Volume 1 [MUR 17a]) whose value depends on the applied voltage when the required frequency range represents a few percentage of the center frequency, or a voltage-to-frequency converter in the case where the variation range must be larger. In both cases, there are maximal and minimal frequencies, $f_{3\max}$ and $f_{3\min}$, on the VCO output, and $f_{2\max}$, $f_{2\min}$ after the divide-by-N frequency divider, respectively, corresponding to the control voltages of the VCO, $V_{\max}$ and $V_{\min}$. Assuming that the frequency–voltage relation is linear, the static transmittance of the VCO is given by $K_f = \dfrac{f_{3\max} - f_{3\min}}{V_{\max} - V_{\min}} = \dfrac{N\left(f_{2\max} - f_{2\min}\right)}{V_{\max} - V_{\min}} = \dfrac{N\,\Delta f_2}{V_{\max} - V_{\min}}$, where Δf_2 is the locking range. The reactivity of the VCO is usually fast enough to consider that the frequency on the VCO output immediately follows the input voltage even for a period of the signal at the loop input. If the maximal sensitivity is adopted for the phase detector, the loop transmittance thus contains the product $\dfrac{K_f}{N}\dfrac{I_0}{C_0} = \dfrac{K_f}{N}\dfrac{I_0 T_e}{C_0 T_e} = \dfrac{\Delta f_2\left(V_{\max} - V_{\min}\right)}{\left(V_{\max} - V_{\min}\right)T_e} = f_2\,\Delta f_2$ which has the dimension of the reciprocal of the squared time, such as $(\omega_n)^2$. It actually shows (see exercise 1.6.4) that the closed-loop angular frequency ω_n is either equal or proportional to $\sqrt{f_2\,\Delta f_2}$ and that if there were no correction filter, the loop transmittance would be proportional to $(\omega_n / s)^2$. A phase-leading correction filter is therefore necessary, otherwise the loop would turn into an oscillator with angular frequency ω_n. In the case where ω_n is equal to $\sqrt{f_2 \Delta f_2}$, which is obtained with a corrective filter using a simple passive circuit implemented by adding a single resistance in series with the capacitor C_0, there is no possible adjustment of ω_n. This is the reason why it is advantageous to introduce an adjustable factor between ω_n and $\sqrt{f_2 \Delta f_2}$ through the presence of at least one correction amplifier in the loop (Figure 1.29), inasmuch as the requirements of the product gain-bandwidth can be properly fulfilled.

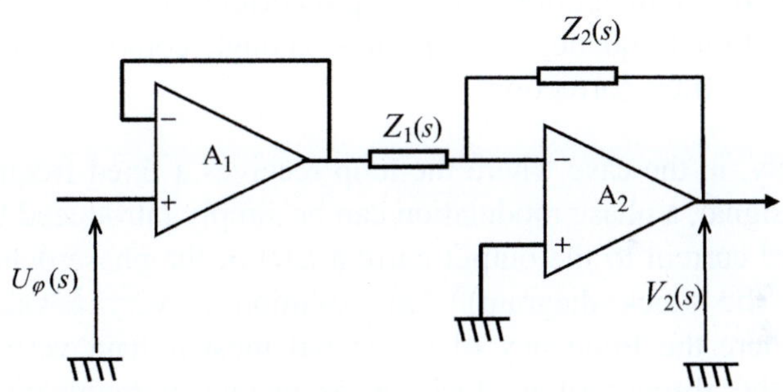

Figure 1.29. *Correction filter allowing for greatest flexibility of adjustment of the parameters of the closed-loop control, consisting of a follower and an inverter with adjustable gain*

1.3.3. *Phase and frequency modulator and demodulator; locking and dynamic operation of the loop*

The demodulation of a phase-modulated signal inserted at the loop input is carried out very easily by making use of the output of the loop filter, which is also the VCO control voltage, and by removing the frequency division (i.e. making $N = 1$). In permanent state, since $H(s)$ is of the second-order low-pass or sum of low-pass and band-pass type, $\lim_{s \to 0}[s H(s)] = 0$, which corresponds to a zero-phase static error. The phase $\varphi_2(t)$ of the signal of the feedback loop is therefore following the phase of the input signal $\varphi_1(t)$, up to the dynamic error, which depends on the cutoff frequency of this second-order system, close to $\omega_n/2\pi$. At the limit of a rather low modulation frequency with respect to this cutoff frequency, the VCO maintains this phase error close to zero. However, given that relations between phase and instantaneous frequency are $\varphi(t) = 2\pi \int_0^t f(t')\,dt'$ and symmetrically $f(t) = \frac{1}{2\pi}\frac{d\varphi(t)}{dt}$, the VCO control voltage is $\frac{f_2(t)}{K_f} = \frac{1}{2\pi K_f}\frac{d\varphi_2(t)}{dt}$. It is therefore possible to recover the phase modulation following an integrator added after the loop filter (but outside of the loop so as not to modify the correction), which is consistent with the previous fundamental relations. A more economical solution implies directly using an image of this current to enable access to phase modulation; it is feasible insofar as the capacitance C_0 plays the role of the integrator of the output current of the phase comparator if the modulation frequency remains low enough compared to the cutoff frequency of the filter correction.

Conversely, in the case where the loop receives a fixed frequency and a fixed phase signal, a phase modulation can be simply introduced by adding a superimposed current to the output current $i_\varphi(t)$ of the phase detector (or its LT $I_\varphi(s)$ in the block diagram). This solution is very advantageous in situations where the frequency of the signal must remain very stable and accurate in the absence of modulation, as in telecommunication systems, because the sinusoidal source function driven by a quartz oscillator can be separated from the modulation function independently implemented. This method allows for both phase and frequency modulation inasmuch as the signal necessary for the second is deduced from that required for the first by integration.

Similar principles can be applied for direct frequency modulation and demodulation. The feedback voltage is recovered directly on the input of the VCO when the loop input signal is modulated in frequency. Concerning the modulation, a voltage perturbation superimposed onto the VCO control voltage is injected. In fact, it is important to note that the closed-loop transmittance is the same for phase and for frequency because $H(s) = \frac{\varphi_2(s)}{\varphi_1(s)} = \frac{s\,\varphi_2(s)}{2\pi} \Big/ \frac{s\,\varphi_1(s)}{2\pi} = \frac{F_2(s)}{F_1(s)}$, which is the ratio of the LT of the phases and the ratio of the LT of the frequencies as well. Consequently, the static frequency error is also zero and the loop dynamically reacts in the same manner to phase and frequency variations.

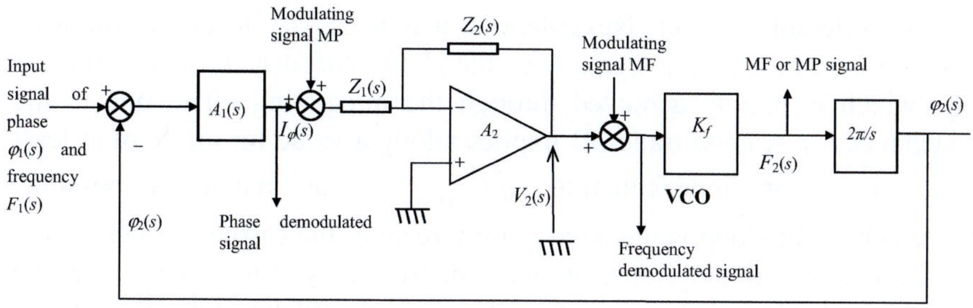

Figure 1.30. *Phase (PM) and frequency (FM) modulators when $\varphi_1(s)$ and $F_1(s)$ are constant; demodulators in the case where $\varphi_1(s)$ and $F_1(s)$ are modulated*

An important aspect of dynamic operation concerns locking and unlocking the loop. When the loop is locked, the shift between phases and instantaneous frequencies of input and output signals is negligible, which corresponds to static operation. However, a perturbation that is fast with regard to the cutoff frequency or brutal such as a phase or frequency step, results in a momentary shift which can be significant and involves unlocking the loop. This occurs especially when the reaction of the loop to this perturbation leads to the saturation at V_{max} and V_{min} of the VCO control voltage and therefore to a nonlinear operation. If the frequency of the input signal remains outside of the locking range $[f_{2max}, f_{2min}]$ accessible to the VCO, it is obvious that the loop cannot be locked again. Otherwise, a very important advantage of the phase detector previously studied is that it also behaves as an all-or-nothing type frequency detector when f_1 and f_2 are different, as already indicated. If $f_1 > f_2$, the control voltage of the VCO filtered by the loop filter shifts to V_{max}, which causes the frequency f_2 to increase on the VCO output and thereby the possibility for this frequency, too low initially, to reach the value of f_1 if the latter is located in the capture range for this state in which the loop was previously unlocked, given that f_1 is already located in the locking range. The consequence is that the capture range is the same as the locking range for this type of phase-shift detector. The dynamics of the feedback in the static state depends on the response of the loop and thus of parameters ω_n (natural frequency) and ζ (damping factor). Their determination must therefore eventually meet the criterion of optimal speed, which is obtained in the neighborhood of $\zeta = \dfrac{1}{\sqrt{2}}$ and a natural frequency $\omega_n/2\pi$ of the order of the maximal modulation frequency

$f_{M\max}$ or preferably greater. Nevertheless, it is necessary to ensure the low-pass filtering of the output signal from the phase-shift detector in an efficient way, which is mainly achieved through the integrating effect due to the capacitance C_0. A good trade-off implies taking a value for $\omega_n/2\pi$ of at least $\sqrt{f_{M\max}\, f_{2\min}}$ or approximating $\sqrt{f_{2\max}\, f_{2\min}}$ as much as possible. Conversely, if the loop is intended to only recover the carrier frequency of a signal modulated in amplitude, in phase or frequency, but comprising a line at this carrier frequency as is often the case, it will be more effective to decrease the natural frequency $\omega_n/2\pi$ and even to increase the damping factor a little in order to reduce the sensitivity of the loop to sidebands due to modulation that can never be completely eliminated by filtering alone. This will be achieved by decreasing the sensitivity of the phase-shift detector, as previously described.

In the case where the PLL is utilized as frequency control, contrary to the previous case, it has to follow abrupt jumps of the input frequency. Its step response is therefore decisive for the agility criterion which characterizes the number of periods after which the output frequency is again equal to that of the input. To decrease this number of periods, we must seek to increase the natural frequency $\omega_n/2\pi$. A very effective solution involves adding an additional correction provided by a derivator to the one including leading phase correction and already implemented (see exercise 1.6.4). Indeed, in this way, the response speed of the loop is improved by the signal derived from $u_\varphi(t)$ which, injected into the input of the VCO, acts to change its frequency in the same direction as the variation of the input frequency. The total closed-loop response then becomes the sum of those of a low-pass with a band-pass filter and it is possible to significantly increase its natural frequency without compromising its stability (see exercise 1.6.4). It results in locking again on the input frequency after a few periods only (<10), which represents a satisfactory performance in terms of agility of the loop. In addition, it provides a means to respond to the imperatives of the demodulations of phase- or frequency-modulated digital signals and of the frequency synthesis examined later in the text. It is nonetheless unwise to derive a signal that has already been integrated, and moreover the use of a differentiator would cause stability issues, as well as problems of gain-bandwidth product of the operational amplifier that is needed to implement this function in addition to an increase of phase noise of the VCO. It is no longer possible to use the current $i_\varphi(t)$, which is actually a series of pulses of duration Δt, which cannot be filtered without introducing a prejudicial delay to the implementation of the desired transmittance for this correction. The solution then involves generating a signal that is directly proportional to the phase difference of the input signals and not to their integral by adding a sample-and-hold circuit that allows for resetting the charge of the capacitance C_0 (Figure 1.31).

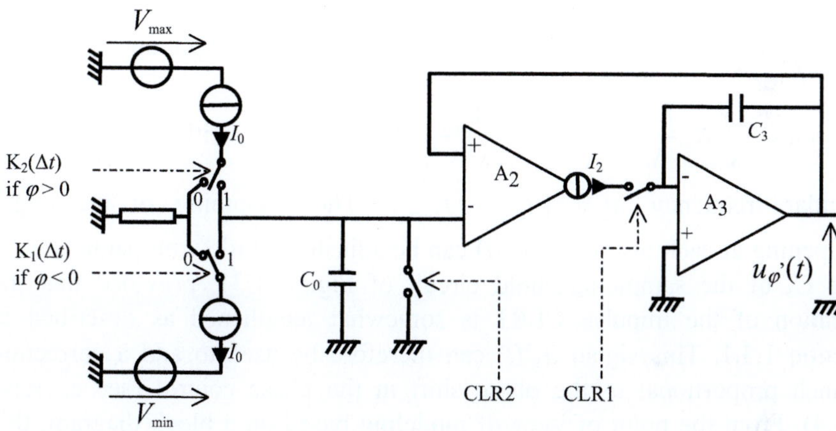

Figure 1.31. *Circuit for measuring the phase shift between square signals; the impulses CLR1 and CLR2, respectively, control the update of the signal after the delay Δt and the zeroing*

To obtain a signal directly proportional to the phase difference $\varphi(kT_e)$ for any period, it suffices to hold its value after the measurement period Δt for the remaining time before the next measurement using the reset impulse of the D flip-flops (/ CLR) of the control circuit (diagram Figure 1.25), which here is renamed CLR1. Next, the charge stored in the capacitance C_0 has to be zeroed by way of short-circuiting before the next measurement. For this last operation, the pulse CLR2 has to be generated, shifted by half a period compared to CLR1, employing a control circuit identical to that of Figure 1.25 but implemented by falling-edge triggered flip-flops or still with the same circuit but receiving the inverted signals $s_1(t)$ and $s_2(t)$. The overall operation may be understood as that of a master–slave analog system based on the alternation of commands CLR1 and CLR2. We thus obtain on the circuit output of Figure 1.31 a signal representing a correct approximation of the derivative $u_\varphi'(t)$ of the signal $u_\varphi(t)$ in the context of the continuous-time approximation, providing a measure of the phase shift $\frac{I_0}{C_0}T_e\frac{\varphi(t)}{2\pi}$ for every period of the input signal. The optimal sensitivity is obtained as previously when the detector delivers a voltage $V_{max} - V_{min}$ for a phase shift of π, that is $\frac{I_0}{C_0}\frac{T_e}{2} = V_{max} - V_{min}$. The transmittance of this phase-shift detector is then $K_\varphi^{'} = \frac{I_0}{C_0}\frac{T_e}{2\pi} = \frac{V_{max} - V_{min}}{\pi}$ and the loop transmittance without correction becomes $\frac{2\pi}{s}K_\varphi^{'}\frac{K_f}{N} = 2K_f\frac{V_{max} - V_{min}}{N s} = 2\frac{f_{2max} - f_{2min}}{s}$ with a characteristic angular frequency $\omega_c = 2\left(f_{2max} - f_{2min}\right)$. The smoothing of the edges happening at each update of $u_\varphi'(t)$ can be adjusted a little with parameters I_2 and C_3 of the sample-and-hold circuit of Figure 1.31, provided that the duration of the impulse CLR1 is somewhat lengthened as described in section 1.3.1. This signal $u_\varphi'(t)$ can therefore be used to add a correction branch proportional to the phase shift in the phase control (see exercise 1.6.4). From the point of view of modeling based on a block diagram, this operation is fully equivalent to that consisting of adding a derived correction following the integral phase detector of Figure 1.30 or of its functional block obtained by LT in Figure 1.31.

It is possible to use only the single circuit of Figure 1.31 to measure the phase shift associated to an integral-proportional correction in the PLL. Nonetheless, the addition of further correction as described previously thus becomes more difficult to be implemented and the advantage of the first detector operating as a frequency comparator, which proves very useful when input frequency excursions exceed the locking range, is lost. The latter holds for a rather modest simplification of the circuitry since two logic control systems (Figure 1.25) remain necessary for the generation of the impulses CLR1 and CLR2 in the case of the detector of Figure 1.31. Finally, it is important to note that in the presence of a single proportional correction following this last phase-shift detector, the loop transmittance would be just $\frac{\omega_c}{s} = \frac{2\left(f_{2max} - f_{2min}\right)}{s}$, and that accordingly the closed-loop transfer function would be of the first-order low-pass type $\frac{1}{1 + s/\omega_c}$ with a cutoff frequency equal to ω_c. This system is not relevant because it is too inefficient in terms of loop gain error and flexibility of the loop.

1.3.4. *Analog frequency synthesis*

Frequency synthesis utilizes some functions, all based on PLLs in locked operation, which ensures that the signals on both inputs of the phase comparator have the same frequency. The input signal is most often provided by a quartz oscillator, providing high stability for its frequency and those deduced therefrom. The other circuits needed are mainly simple or specialized mixers, low-pass or bandpass filters and simple or specialized frequency divider-counters. Due to cutoff filters with steep slopes, the spectral purity of the sinusoidal output signal is very good, with a rejection of unwanted frequencies that can reach or exceed 100 dB.

1.3.4.1. *Multiplication of a frequency by a fixed or variable fractional number*

By including a PLL equipped with a divide-by-N divider (N integer) after a divide-by-P (P integer) frequency divider in the loop, a frequency N times greater than on the loop input is obtained on the VCO output of the latter, where it has already been divided by P. The resulting operation indeed consists of the product of the input frequency by N/P. The function will be symbolized by a rectangle containing the ratio N/P as in Figure 1.32.

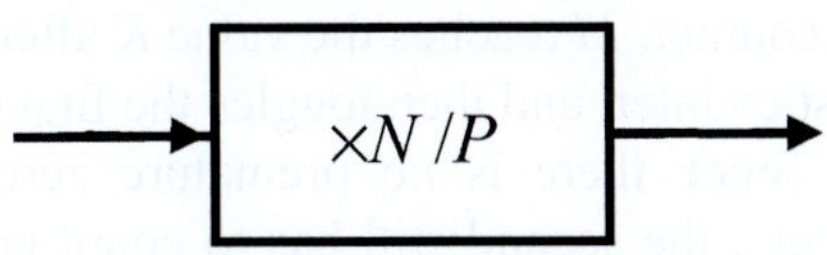

Figure 1.32. *Frequency multiplier using multiplication ratio N/P*

The implementation can be done conventionally based on counters whose zero resetting after N or P clock ticks results from the detection of the numbers N or P using a combinational logic system when the numbers N and P are constant. On the other hand, if one of these numbers must vary by more than 2 units, which occurs very often in frequency synthesis, it is advantageous to resort to a split counter using division rate switching. A first counter divides by $P1$ or $P2$ depending on the value 0 or 1 of a control bit V, which only requires a detection system in which a 2-to-1 multiplexer controlled by V has been added to perform the zero reset command. A second counter placed in cascade after the first one divides by a fixed integer $P3$. A third circuit carries out the bit-to-bit logic comparison of the word M transmitted to the parallel outputs of the second counter with the instruction word K ($<P3$) that represents the desired variation for the overall counting rate.

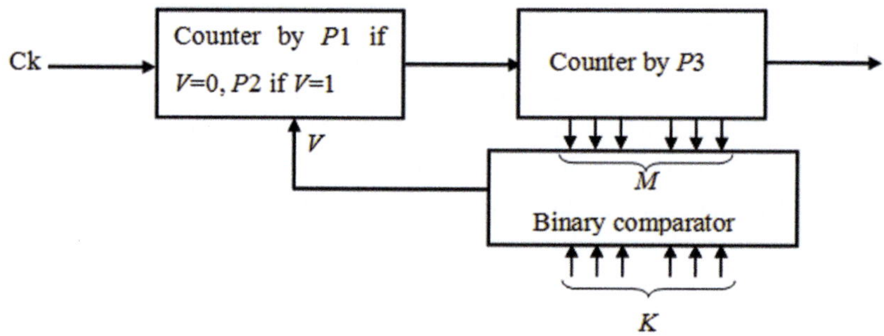

Figure 1.33. *Divide-by-P2× P3 + K× (P1 – P2) frequency*

It outputs $V = 0$ if $M < K$ and $V = 1$ if $M \geq K$. The first of both conditions is implemented at the start of the counting and accordingly the first counter divides by $P1$, which drives the second counter into increasing its outputs by one only after $P1$ clock ticks on the first counter input. Therefore, on the outputs of the second counter, M reaches the value K after $P1 \times K$ clock ticks on the input of the first counter, and then toggles the first counter to shift to a division by $P2$ rate. Since there is no premature zero reset due to any circuitry for the counters, the second still has to count up to $P3$ to return to its original state and to this end the first still has to receive $P2 \times (P3 - K)$ clock ticks. Subsequently, $P1 \times K + P2 \times (P3 - K)$ clock ticks are necessary on input in total, or still $P2 \times P3 + K \times (P1 - P2)$.

Given that it is often desirable that this result be increased by one unity as for K, it suffices to take $P1 - P2 = 1$, and the other values are determined by the limits that the overall division rate should assume, which can easily exceed 100, and from practical considerations concerning every counter.

Details of the frequency divider circuit are given in Figure 1.33, while the division function and the multiplication function obtained by inserting the divider in a PLL are symbolized in Figure 1.34.

Figure 1.34. *Frequency multiplier on the left and frequency divider on the right, each with unit step for the control number K*

Exercise: determine the parameters to obtain a count from 290 to 299.

Answer: If K varies from 0 to 9, we must have $P2 \times P3 = 290$, which is not divisible by a power of 2 except 2. However, if K varies from 2 to 11, $P2 \times P3 = 288$, which is divisible by 32. There is thus interest in choosing this solution with $P3 = 32$, $P1 = 10$ and $P2 = 9$, which implies no combinatorial network or at most an AND gate for the resetting detection.

1.3.4.2. *Frequency shift*

It is often useful in many systems, especially in telecommunications, to shift an initial frequency f_1 of a fixed value or a variable Δf. The simple method previously described which consists of adding a voltage modulation on the VCO input of the PLL may be appropriate if there are no concerns about the loss of accuracy in the value of the frequency which is not any longer only driven by a quartz oscillator. Nonetheless, this disadvantage is generally not tolerable in frequency synthesis.

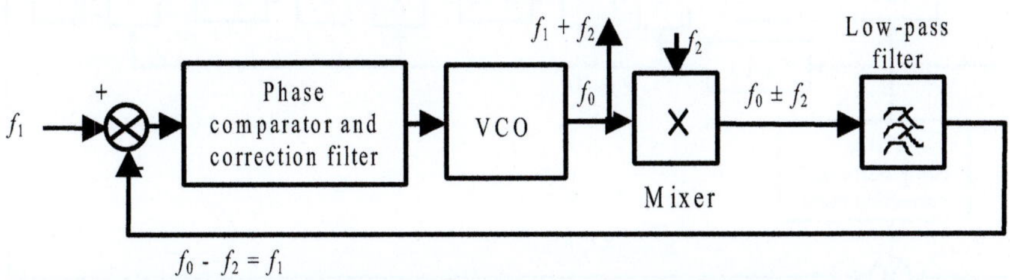

Figure 1.35. *Frequency-shift loop based on the locking condition* $f_1 = f_0 - f_2$ *that yields an output VCO frequency* $f_0 = f_1 + f_2$

It is however possible to preserve the advantage from the stability provided by the quartz clock-based control by inserting after the VCO a mixer that achieves the product of the output signal of the VCO at frequency f_0 by an external signal at frequency f_2, followed by a low-pass filter that provides a signal frequency equal to the difference of the frequencies existing on the input of the mixer, namely $f_0 - f_2$ (Figure 1.35). Since the loop is locked, the low-pass filter output must be achieved at frequency f_1 because it is transmitted back to the input through the feedback loop (it is assumed that there is no division of frequency inside the loop). Consequently, the output frequency of the VCO must be $f_0 = f_1 + f_2$ so that $f_1 = f_0 - f_2$ and it suffices to choose $\Delta f = f_2$, this last frequency being controlled by a quartz oscillator if necessary.

1.3.4.3. *Multiplication of a shift between two frequencies*

A slightly more complex system than the previous one allows us to make a conversion of frequency using a multiplicative factor which affects the shift Δf_1 between the two input frequencies of the system, f_0 and f_1, close enough, in other words differing at most by a few percentage. It comprises a main loop in which several operations carried out by frequency division and/or multiplication as well as mixer assemblies added to a bandpass filter centered on f_0 are necessary. The block diagram is given in Figure 1.36 for a multiplication factor of 1,000 in the case where $f_1 = f_0 + \Delta f_1$. The VCO provides the frequency $f_2 = f_0 + \Delta f_2$ in which the shift Δf_2 is to be determined from the requirement for the lock of the overall loop.

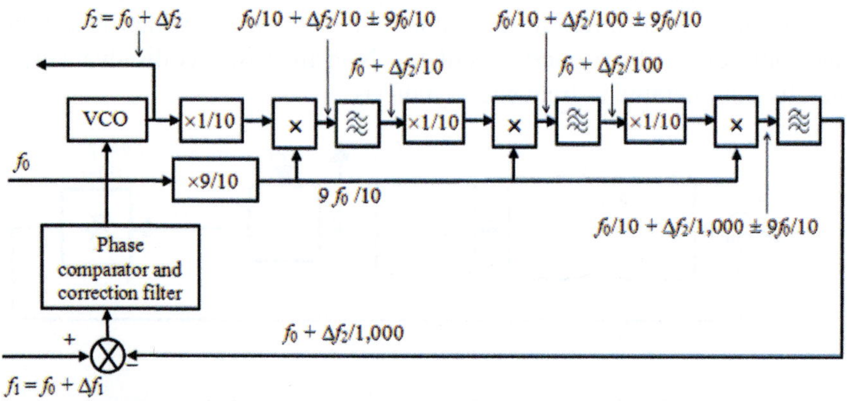

Figure 1.36. *Frequency shift multiply-by-1 000 multiplication loop*

The succession of divide-by-10 divider, mixer with the signal of frequency $9f_0/10$ and bandpass filter blocks implement a circuit that reaches a frequency of $f_0 + \Delta f_2/1000$ in the feedback loop. The lock thus imposes $f_0 + \Delta f_1 = f_0 + \Delta f_2/1000$, that is $\Delta f_2 = 1000\ \Delta f_1$. The output frequency f_2 is thus shifted by an offset 1,000 times larger compared to f_0 than the input frequency f_1 was because of the three stages of division by 10, mixing (allowing the beat to be obtained) and bandpass filtering inserted in the loop. By adding a mixer outside of the loop with the frequency signal f_0 on the output of the VCO followed by a low-pass filter, it is then very easy to obtain the signal of frequency 1,000 Δf_1.

An important application of this frequency-shift multiplier can be found in fast and highly accurate frequency meters. The basic operation of a frequency meter ensues from counting the number of periods of the signal which we want to measure the frequency of, during a fixed duration T_1 determined very precisely by a quartz clock and a counter-divider. When it is desirable to obtain a measurement to the hertz, we must take T_1 = 1s. If we want an precision to the thousandth of hertz, it would be necessary that T_1 = 1,000s, which is a very long time. In addition, this would imply that the phenomenon whose frequency is being measured be perfectly stationary during this period. The solution consists of making the first measurement of the number of periods with a counter counting for a duration T_1 = 1s to obtain a number M equal to the frequency f_0 with a precision to 1 hertz. This number is immediately used in a frequency synthesizer (which can be regarded as a PLL equipped with a down-counter with parallel load of the number M) to generate a signal of frequency f_0 with three zeros after the decimal point, to the nearest thousandth of hertz, which requires a precision quartz clock and adequate stability. However, the actual signal has a frequency f_1 which may differ from f_0 with a shift that can range from 0 to 999 mHz. Using the multiply-by-1,000 shift multiplier system, a new frequency shift is obtained between 0 and 999 Hz that can be measured in turn in 1s. Furthermore, this measure will provide three decimals after the number M during a period 1,000 times smaller than with a basic frequency meter.

1.3.4.4. *Frequency synthesizer*

Actual frequency synthesis consists of the construction of the frequency of a signal, usually in the decimal number system, by choosing every digit, corresponding for example to the hundredth of hertz, tenths of hertz, unity, ten, hundred, thousand hertz, tens of kilohertz and so on. Several systems are possible but are generally based on a subsystem that can be cascaded as many times as there are digits to be defined and numbered starting from $i = 0$. This subsystem can be called "digital insertion unit" (or "decimal insertion unit") and it will constitute the basic functional block of the synthesizer. It will be necessary to add a quartz clock and a divide-by-M divider, defining the two frequencies f_0 and f_0/M all the more accurately that there will be significant digits to be programmed for the frequency of the output signal. The frequency f_0 should be at least twice as high as the maximal output frequency of the synthesizer. The number M is such that the frequency f_0/M is equal to the frequency corresponding to the most significant digit of the synthesizer output frequency when it is set to the value 1.

This decimal insertion unit, which uses a frequency multiplier based on a split counter as described in section 1.3.4.1, is studied first, and is pictured in Figure 1.37 in which frequencies are indicated, with the assumption that the loop is locked. Its function is to shift the input frequency with an offset equal to the product of an integer number K_i that is chosen between 0 and 9, with the frequency f_0/M and to divide the previous shift by 10. If the input frequency of the digital insertion unit is $f_0 + \Delta f_i$, the new shift from f_0 with respect to the output becomes $\Delta f_{i+1} = \Delta f_i/10 + K_i f_0/M$, with $\Delta f_0 = 0$, which means that the first unit receives the frequency f_0 and for the following units the previous shift is divided by 10.

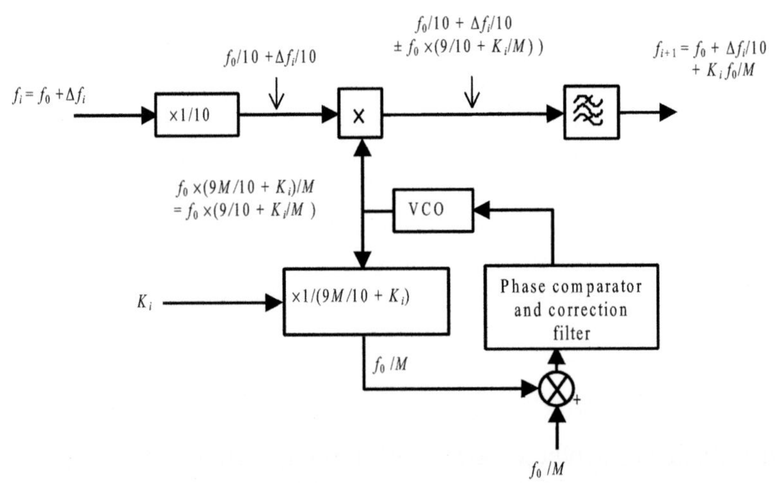

Figure 1.37. *Digital (or decimal) insertion unit*

After cascading n+1 units, it is deduced that the output frequency is $f_0 + (K_0 + K_1/10 + K_2/100 + K_3/1000 + ... + K_n/10^n) f_0/M$ which will be sent into a mixer with the frequency f_0 to recover after low-pass filtering the frequency difference corresponding to all of the terms of the previous sum except the first one.

If, for example, we want to reach 999,999 Hz with a resolution of 1 Hz, by taking f_0 = 10 MHz, it will be necessary that f_0/M = 100 kHz, that is M = 100, and the number measuring the output frequency will consist of digits $K_0\, K_1\, K_2\, K_3\, K_4\, K_5$ from left to right (K_0 being the most significant, K_5 being the digit of the units). In the loop of each unit, the divider based on a split counter will have to divide by $90 + K_i$.

1.3.5. *Digital synthesis and phase and frequency control systems*

In order to design purely digital frequency synthesis systems, certain devices used in analog synthesis have to be transformed in such a manner that the flexibility is removed and limitations are imposed on a number of performances. The systems thus obtained however are much less expensive and easier to incorporate within an integrated circuit composed exclusively of logical functions. Only the direct digital synthesis of a sinusoidal signal and the digital PLL are discussed. The study of digital loops is based on the properties of the type 297 circuit from Texas Instruments [HC 297], initially built using TTL LS technology (Low Power Schottky), and then using CMOS technology, namely HC and HCT series.

1.3.5.1. *Digital frequency synthesis*

A simple way to generate a sinusoidal signal with predetermined frequency relies on the storage in a read-only-memory (ROM) of the digital values $\sin(2\pi i/N)$ of a sine wave sampled with N points for a period, numbered from $i = 0$ to $i = N-1$. It then suffices to read and transmit to an analog-to-digital converter a number $n \leq N$ of these values evenly spaced at the pace of a fixed sampling period T_e. The period of this signal, illustrated in Figure 1.38 for n = 16, for example, will thus be nT_e and it will be possible to adjust its frequency $1/(nT_e)$ using the number n. When $n = N$, all samples are utilized and the frequency reaches its minimal value equal to $1/(NT_e)$. For $n < N$, an integer number p should in theory still exist such that

$n \times p = N$ so that the repeated signal be periodic, or, in other words, that the remainder of the division N/n be always zero. In practice, we jump $p - 1$ values between each sample used, making a total of $n \times (p - 1) + n = n \times p$ which should be equal to the total number N of values available in the ROM and the number i should be a multiple of p. This rule may not always be verified and it should be relaxed by tolerating a fluctuation in the number of unused samples between each chosen sample, which amounts to adjusting the number i to the closest value of a multiple of the elementary phase shift $2\pi/n$ for the chosen frequency $1/(nT_e)$. A distortion results thereof, all the less significant that the number N is larger, which adds up to the one resulting from the shape of the visible step-shaped signal in Figure 1.38. Concerning the latter, by calculating the Fourier series decomposition it can be shown that in addition to the fundamental line at frequency $1/(nT_e)$, other lines can be found at larger frequencies $(n \pm 1)$ and their multiples, namely for the lowest, at frequencies very close to $1/T_e$. It is therefore sufficient to add to the output of the digital-to-analog converter a low-pass filter at a fixed cutoff frequency close to $1/(2T_e)$. Bearing in mind that the first distortion is similar to amplitude modulation with period T_e or smaller, thereby giving lines at frequencies $1/T_e$ or higher and their multiples, the low-pass filter is effective in attenuating the two types of default and restore a sinusoidal signal of acceptable spectral purity. Provided that a microcontroller be employed to manage the choice of the sample numbers, this solution proves to be quite economical if tolerating a frequency adjustment step that is not constant. Other systems do exist, such as the one based on cascading digital rate multipliers (or DRMs), which are counters providing an adjustable number of impulses selected among those present on the input. The output signal then contains phase noise that can only be reduced by increasing the frequency division added after the last DRM. This problem also arises in the digital loop studied hereafter. Generally speaking, these systems exhibit more limited performance but have the advantage of requiring less analog components and as a result, they are less expensive and integrate more easily in assemblies where logical components predominate.

Exercise: if $N = 2^{16}$, what are the numbers k of samples giving the closest amplitudes to those of the ideal sinusoid for $n = 15$ samples in total? What is the ratio between the corresponding frequency and the frequency $f_{\min} = 1/(2^{16}T_e)$ obtained when all samples are used?

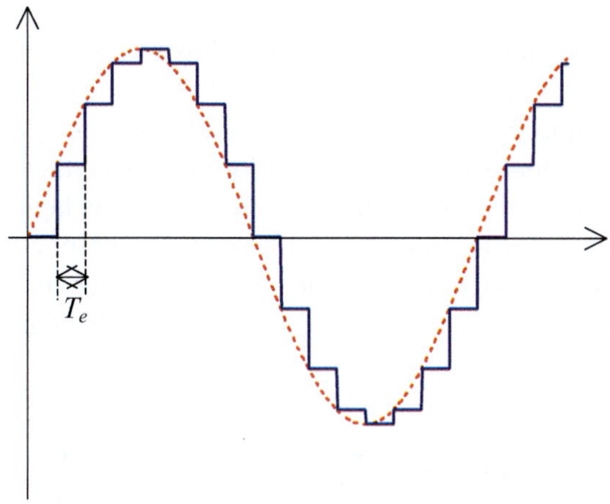

Figure 1.38. *Output signal of the converter (solid line) in the case of 16 samples per period*

Answer: for $i = 0$ to 14, the phases are equal to $2\pi \times i/15$, whereas values corresponding to the phases $2\pi \times k/2^{16}$ are available. It is thus necessary to find fractions $i/15$ the closest to $k/2^{16}$ by rounding to the nearest to $2^{16} \times i/15 = i \times 4369.0666667$. We will therefore round down to the next greatest integer from $i = 0$ to 7 and up to the next smallest integer from $i = 8$ to 14. The corresponding frequency is $1/(15T_e)$, that is $2^{16} f_{\min}/15 = 4369{,}0666667\ f_{\min}$.

1.3.5.2. *Digital PLL*

A purely digital PLL must comprise a phase-shift detector, which can be either an exclusive-OR or an edge-triggered RST (or RSh) flip-flop, which will be taken into consideration here, as well as a logical system capable of delivering a variable frequency signal according to the measured phase shift, intended to replace the VCO [HC 297].

When the loop is locked and in a stationary state, signals at the inputs S ("set") and R ("reset") of the falling-edge triggered flip-flop, identified by their frequencies f_{in} and f_{out}, are shifted of half a period, which results in the output Down-Up (DU) of the flip-flop being a signal with a duty cycle of 1/2 (Figure 1.39). The phase shift is then considered as zero because the *DU* bit commands either a counting up or a counting down action, which leaves the system in a steady state if these two actions last the same time. If we deviate from this state of equilibrium by shifting the falling edges by $T_{in}\,\Delta\varphi\,/(2\pi)$ as in Figure 1.39, the output signal of the flip-flop features either a smaller time duration of the state H when the signal at frequency f_{out} is lagging behind the signal at frequency f_{in}, or a larger one in the opposite case, compared to the half-period $T_{in}/2$ of the signal at frequency f_{in}. In order to ensure control feedback, the state L of *DU* will be used to accelerate the signal f_{out}, and the state H of *DU* to slow it down.

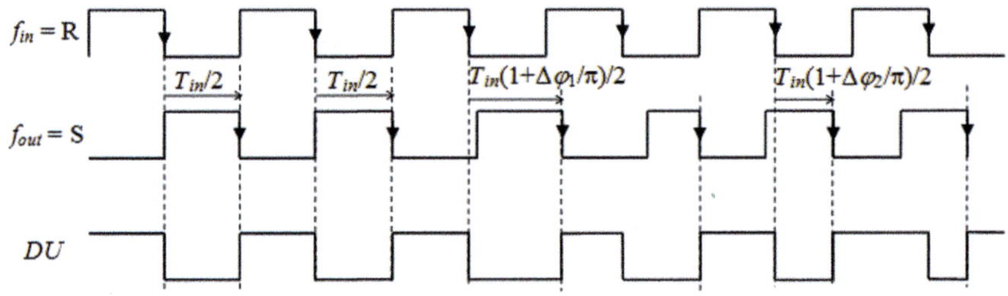

Figure 1.39. *Input signals of the phase-shift detector using RS falling-edge triggered flip-flops and DU output, with zero phase shift over the first two periods, then positive $\Delta\varphi_1$ corresponding to a delay of the signal at frequency f_{out} (or a lead of the signal at frequency f_{in}) in the middle, then the opposite for $\Delta\varphi_2$ (negative) at the end of the sequence*

For the system functioning as VCO, it is essential that it be driven by an external clock working at a much higher frequency than the frequency range of the input signal in order to make it possible to obtain, after division, a variation as regular as possible of the frequency and the signal phase in the feedback loop, despite the impossibility of obtaining a continuous variation as in an analog VCO. This system is based on coupling an up-down counter (CDK in Figure 1.40), having a count rate equal to *K*, and delivering *C* carry or *B* borrow bit once the up-count or down-count has respectively reached the value *K* or zero, with a divide-by-2 divider capable of adding an impulse or of removing one according to the state of the logical variables *C* and *B* which are received as active falling-edge inputs. For the first CDK unit, the up-counter or down-counter function is activated according to the status of the *DU* bit received from the phase detector. With a clock signal having a duty cycle of 1/2, the impulse adder or remover counter (IARC) delivers one impulse in two in a regular fashion when *B* and *C* are not active but respectively adds one or removes one thereof when the edge of C or B falls (Figure 1.40). This occurs when the state of one of the C or B carry bits on the output of the first up-down counter returns to L one clock period after the count rate has reached K, or when the countdown rate has reached zero (Figure 1.40). To guarantee the feedback, DU must therefore activate the up-count for state L and the down-count for state H.

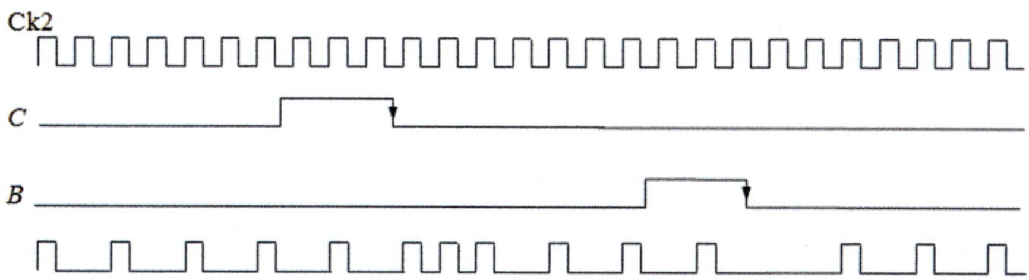

Figure 1.40. *Clock signals Ck2, carry C, borrow B and output of the IARC, from top to bottom*

In addition, the full loop (Figure 1.41) includes a divide-by-N divider inserted on the output of the IARC as well as the dividers necessary to deliver adequate clock frequencies, referenced to the central frequency f_0 imposed by the control frequency $2MNf_0$.

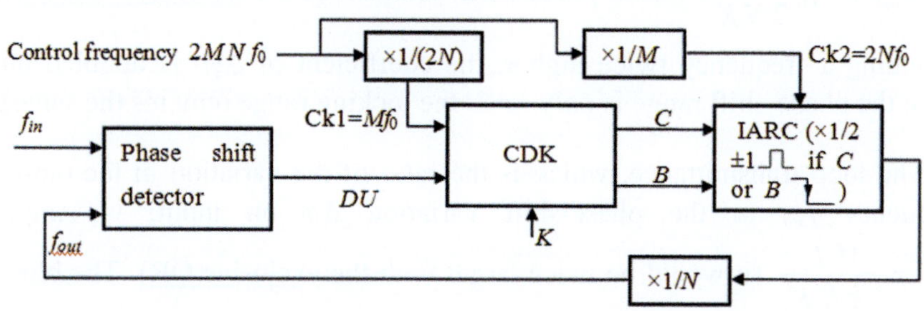

Figure 1.41. *Schematic of the digital phase-locked loop*

By evaluating over one period separating two successive edges with the same DU direction, it can be considered that the CDK counter delivers a falling edge on C or B every time the number of clock ticks Ck1 (as many as Mf0 per second) exceeds K during the time period $T_{in}\,\Delta\varphi/(2\pi)$ (see Figure 1.41) equal to the difference of times spent at L and H, or a number of ticks $M f_0 T_{in} \Delta\varphi/(2\pi K) = M f_0 \Delta\varphi/(2\pi K f_{in})$. The IARC then increases or decreases by one unity the number of impulses that it outputs every time it receives a falling edge on C or B according to the sign of $\Delta\varphi$ during the period of the DU signal, which is equal to $1/f_{in}$. The natural frequency on output of the IARC in the absence of transition on C or B is Nf_0, that is to say, number of ticks Nf_0/f_{in} during the period of the DU signal. The number of ticks therefore becomes $\frac{Nf_0}{f_{in}} + \frac{Mf_0 \Delta\varphi}{2\pi K f_{in}} = \frac{f_0}{f_{in}}\left[N + \frac{M \Delta\varphi}{2\pi K}\right]$, which is an average frequency $f_0\left[N + \frac{M \Delta\varphi}{2\pi K}\right]$ during the period of the DU signal. Nonetheless, this is an average frequency since the rhythm of the output impulses is not always $1/Nf_0$, because the impulses added or removed correspond to a phase shift, respectively, decreased or increased by π. This is why it is paramount to divide once more the frequency by N for this phase fluctuation (or jitter) to be reduced to $\pm\pi/N$. Finally, the average output frequency of the divide-by-N divider becomes $f_{out} = f_0\left[1 + \frac{M \Delta\varphi}{2\pi N K}\right]$. The central frequency can be deduced therefrom when $\Delta\varphi = 0$, which is f_0. And from the maximal variation $\pm\pi$ of $\Delta\varphi$, the locking range is calculated, which is $\Delta f_{out} = \pm f_0 \frac{M}{2NK}$ (for a phase-shift detector of the exclusive OR-type, providing a frequency twice higher, the coefficient of $\Delta\varphi$ is doubled but since the phase-shift range is only $\pm\pi/2$, the locking range remains the same).

The loop transmittance, which is the ratio of the variation in the output frequency f_{out} to the phase-shift variation $\Delta\varphi$ on input, is simply $T_{dgtl} = \frac{M f_0}{2\pi N K}$ (it would be twice larger with the exclusive OR). The block diagram (Figure 1.42) can be deduced thereof as well as closed-loop transmittance, which is of the first-order low-pass type due to the presence of the transmittance $2\pi/s$ that transforms the LTs of the frequency into phase LTs, similarly to the analog loop. The closed-loop transmittance is thus $\frac{2\pi T_{dgtl}/s}{1 + 2\pi T_{dgtl}/s} = \frac{1}{1 + \frac{NK}{M f_0} s}$ with a cutoff angular frequency $\omega_c = \frac{M f_0}{NK}$.

Choice of parameters and phase noise reduction

The division rates M, N and K are determined based on criteria concerning the cutoff frequency $\omega_c = \dfrac{M f_0}{N K}$, the locking range $f_{out} = f_0\left[1+\dfrac{M\,\Delta\varphi}{2\pi N K}\right]$, whereas the central frequency of the locking range is set by f_0. The phase noise is characterized by a phase jitter $\pm\pi/N$. It is necessary that M/K must be kept well below N to ensure a pace of the edges on C and B slower than Ck2 to decrease the phase jitter. In order to go further, N must be taken as large enough, which requires M to be set to an even larger value.

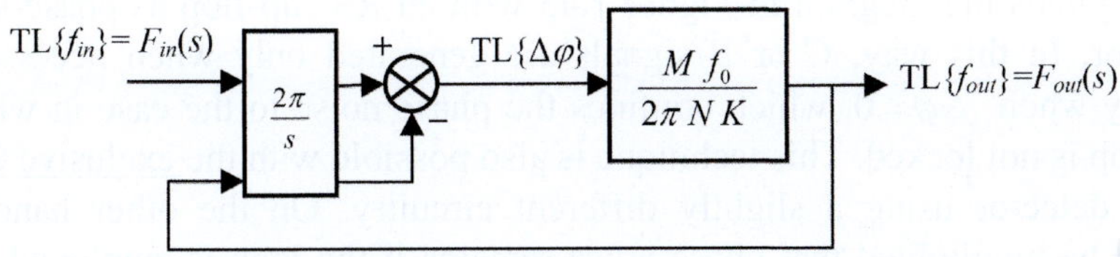

Figure 1.42. *Block diagram of the digital PLL*

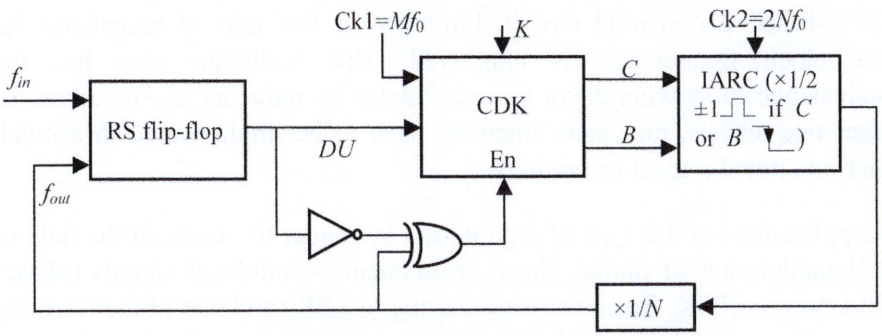

Figure 1.43. *Digital PLL with phase noise reduction*

Nevertheless, it is possible to remove this phase noise in steady state when the loop is locked with $f_{out} = f_{in}$, close to f_0. In effect, if the number of clock ticks Ck1 is enough for half a period of the DU signal, that is to say also half a period of input and output signals $1/(2f_{out})$, the up-counts or down-counts achieved by the CDK counter will overflow the values K or 0, thus generating addition and removal of impulses by the IARC for each period of the output signal. The two effects will cancel each other at the level of the average frequency but lead to unnecessary phase fluctuations that can possibly be removed by adjusting the count rates in such a way that these overflows do not occur, that is to say $M f_0 /(2f_0) < K$, or $K > M/2$.

However, a more radical solution involves activating the CDK counter only when strictly necessary, namely when $\Delta\varphi \neq 0$. To do this, we utilize a validation input "En" ("Enable") of this counter that requires it to remain at rest, while $\Delta\varphi = 0$. Provided that we have a duty cycle of 1/2 for signal f_{in} and that N is chosen as being even, which causes that the duty cycle of the signal f_{out} is also 1/2, this signal can be generated by carrying out the exclusive OR of the complement of the DU signal with the output signal f_{out}, which yields the diagram in Figure 1.43 with an RS flip-flop as phase-shift detector. In this way, C or B signals are generated only when necessary, namely when $\Delta\varphi \neq 0$, which confines the phase noise to the case in which the loop is not locked. This technique is also possible with the exclusive OR-based detector using a slightly different circuitry. On the other hand, it should be highlighted that phase noise remains if the loop is employed, for example, to demodulate a phase or frequency modulated signal. The only way to reduce it in this case is therefore to increase N, but this leads to an increase in clock frequencies which can limit f_0 to a few megahertz if the HC297-based commercial circuit, limited to a few tens of megahertz for these clock frequencies, is employed. This technique also has the disadvantage of slowing down transient states by reducing ω_c. To allow for higher frequencies, the same functions have to be implemented in a much faster integrated logical technology.

Applications of this type of digital loop fall under the scope of the field of the demodulation of digital phase or frequency-modulated signals (phase-shift keying or PSK, frequency-shift keying or FSK) and that of the recovery of carriers, useful, for example, for read accesses in hard drives. Complementary circuits enabling locking detection, the extension of the locking range by splitting the divide-by-N divider in two, the use of the parallel outputs of the divide-by-N divider, parameter control using microprocessors and so forth, can be added in order to increase the versatility of the loop and to adapt it to the requirements of the intended application.

Digital second-order PLL

The digital first-order PLL previously studied has the disadvantage of having a non-zero phase error from the moment the frequency $f_{in} = f_{out}$ deviates from f_0, precisely because of the first-order response, and consequently incompressible residual phase noise. Therefore, seeking an implementation of a second-order closed-loop transmittance would prove to be useful but this is not possible by fiddling with a corrector as in an analog loop because there is none! As a result, it is necessary to entangle two digital loops, for example, in the following fashion, which leads to quite a complex system (Figure 1.44).

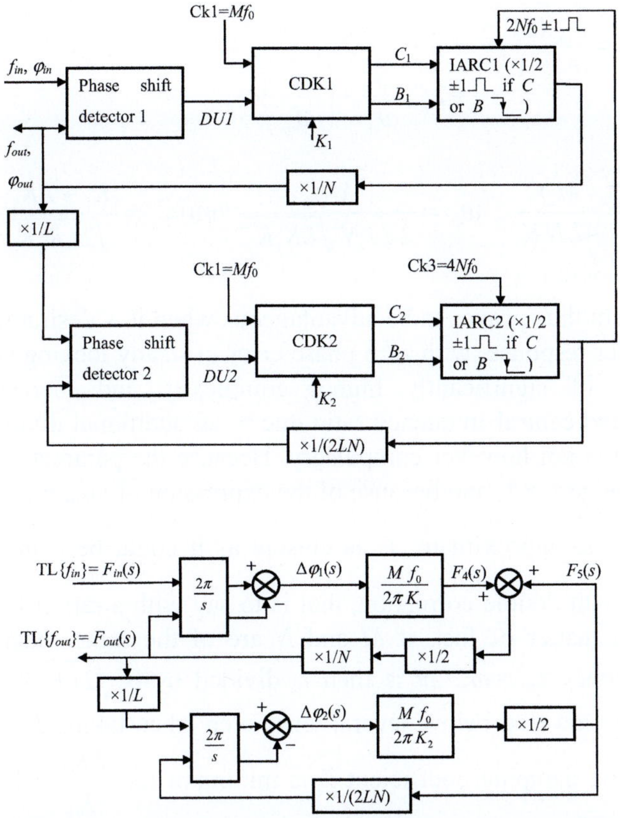

Figure 1.44. *Circuit assembly of a digital second-order PLL (on top) and its operating block diagram (at the bottom)*

From Figure 1.44, we deduce $F_4(s)=\frac{2\pi}{s}\frac{M f_0}{2\pi K_1}\left[F_{in}(s)-F_{out}(s)\right]$; $F_5(s)=\frac{2\pi}{s}\frac{M f_0}{4\pi K_2}\left[\frac{F_{out}(s)}{L}-\frac{F_5(s)}{2LN}\right]$ and $F_{out}(s)=\frac{1}{2N}\left[F_4(s)+F_5(s)\right]$.

By eliminating $F_4(s)$ and $F_5(s)$ in these three equations, a second-order closed-loop transmittance is obtained of the same nature as that of the analog loop with double correction, summing a low-pass and a bandpass filter transmittance:

$$H_n(s)=\frac{F_{out}(s)}{F_{in}(s)}=\frac{1+\dfrac{s}{\omega_{bn}}}{1+2\zeta_n\dfrac{s}{\omega_{nn}}+\left(\dfrac{s}{\omega_{nn}}\right)^2}$$

with $\omega_{bn}=\frac{M f_0}{4LNK_2}$, $\omega_{nn}=\frac{M f_0}{2\sqrt{2}N\sqrt{LK_1K_2}}$ and $\zeta_n=\frac{K_1+LK_2}{\sqrt{2LK_1K_2}}$.

This system thus proves to be advantageous when it is desirable to obtain a second-order response, with zero phase error in steady locking state, but at the expense of significantly higher complexity and increased clock frequencies and central frequency ratio due to an additional divider (divide-by-L), which is not however compulsory. Because the parameters *M, N*, K_1 and K_2 are integers > 1, and because of the expression of ω_{nn}, it is impossible for $f_{nn}=\frac{\omega_{nn}}{2\pi}$ to approximate f_0 as closely as it could be achieved in the analog loop with double correction, that is to say with a ratio of only a few unities. As a matter of fact, if *M* and *N* are of the same magnitude, the natural frequency $f_{nn}=\omega_{nn}/2\pi$ is then f_0 divided by $4\pi\sqrt{2LK_1K_2}$, a factor that cannot be lowered below several tens, even when taking $L = 1$. On the other hand, the damping coefficient ζ_n is minimum for $\frac{LK_2}{K_1}=1$ and is in this case equal to $\sqrt{2}$. Consequently, it is impossible to obtain a damped oscillating step response. These two limitations thus make this loop poorly flexible and its only advantage involves a zero phase error in static mode. In conclusion, the technological benefit of its utilization, due to the purely digital nature of the circuits employed and the possibility of controlling its parameters with a microcontroller, must be compared to the inherent drawbacks, to the limitations in terms of frequency and to those imposed on the parameters of the control.

1.4. Sampled systems

1.4.1. *Z-transform for systems described by a recurrence equation (or difference equation)*

We shift from the LT of the response $y(t)$ of a continuous-time system to that of the corresponding sampled system by applying the "Dirac comb" distribution with time period T_e on $y(t)\exp(-st)$:

$$Y(z)=\int_0^{\infty}\sum_{k=-\infty}^{\infty}\delta(t-kT_e)\,y(t)\exp(-st)\,dt=\sum_{k=0}^{\infty}y(kT_e)\exp(-skT_e)=\sum_{k=0}^{\infty}y[k]\,z^{-k}.$$

The *z-transform* (ZT) is thus defined by establishing $\boxed{z=\exp(sT_e)}$ and $\boxed{z^{-1}=\exp(-sT_e)}$ corresponding to the delay operator of one sampling period. Denoting the samples by $y[k]$ to shorten the notation $y(kT_e)$ and to use the kth-order number of the sequence, we can write:

$$ZT\{y(t)\}=\sum_{k=0}^{\infty}y[k]\,z^{-k}.$$

The ZT of a recurrence equation or difference equation exhibiting a linear combination of the different samples $y[i]$, $y[i-1]$, $y[i-2]$, and so on, is thus easily obtained by multiplying the value of each sample by z^{-i}, z^{-i+1}, z^{-i+2} and so on, according to the delay of the sample with regard to the time origin, measured in the sampling period unit T_e. The variable z being dimensionless, <u>the ZT retains the dimension of the function $y(t)$</u>.

The resulting series is convergent provided that $|z| > L$, in which L is a convergence radius. From the identity $\sum_{i=0}^{n}u^i=\frac{1-u^{n+1}}{1-u}$, it is deduced that the geometric series $\sum_{i=0}^{n}(az)^{-i}=\frac{1-(az)^{-n-1}}{1-(az)^{-1}}$ converges to $\frac{1}{1-(az)^{-1}}$ when $n\to\infty$ only if $|az|>1$, that is if $|z|>\frac{1}{|a|}$, hence $L = a$. We can then calculate the transform of all ordinary functions with the Heaviside step function U(t) in factors to ensure the causal nature, which allows us to create the dictionary of transforms hereafter (Table 1.1).

$y(t)$	$y[k]$	$Y(z) = ZT\{y(t)\}$
$\delta(t)$	$y[0] = 1/T_e{}^*$ and $y[k] = 0$ for $k \neq 0$	$1/T_e{}^*$
$\delta(t-kT_e)$	$y[k] = 1/T_e{}^*$ and $y[k'] = 0$ for $k' \neq k$ * see below and note in section 7.4.2	$z^{-k}/T_e{}^*$ * see below and note in section 7.4.2
U(t)	$y[k] = 1 \ \forall k \geq 0$	$\frac{1}{1-z^{-1}}$
t U(t)	$y[k] = k \ \forall k \geq 0$	$\frac{z^{-1}}{(1-z^{-1})^2}$
$\exp(-\alpha t)$ U(t) with $\alpha > 0$	$y[k] = \exp(-\alpha kT_e) \ \forall k \geq 0$	$\frac{1}{1-e^{-\alpha T_e}z^{-1}}$
$\exp(-\sigma t)\cos(\omega_0 t)$U($t$) with $\sigma > 0$	$y[k] = \exp(-\sigma kT_e)\cos(\omega_0 kT_e)$ $\forall k \geq 0$	$\frac{1-e^{-\sigma T_e}z^{-1}\cos(\omega_0 T_e)}{1-2e^{-\sigma T_e}z^{-1}\cos(\omega_0 T_e)+e^{-2\sigma T_e}z^{-2}}$
$\exp(-\sigma t)\sin(\omega_0 t)$U($t$) with $\sigma > 0$	$y[k] = \exp(-\sigma kT_e)\sin(\omega_0 kT_e)$ $\forall k \geq 0$	$\frac{e^{-\sigma T_e}z^{-1}\sin(\omega_0 T_e)}{1-2e^{-\sigma T_e}z^{-1}\cos(\omega_0 T_e)+e^{-2\sigma T_e}z^{-2}}$
$\exp(-\sigma t)\cos(\omega_0 t+\varphi_0)$U($t$) with $\sigma > 0$	$y[k] = \exp(-\sigma kT_e)\cos(\omega_0 kT_e+\varphi_0)$ $\forall k \geq 0$	$\frac{\cos(\varphi_0)-e^{-\sigma T_e}z^{-1}\cos(\omega_0 T_e-\varphi_0)}{1-2e^{-\sigma T_e}z^{-1}\cos(\omega_0 T_e)+e^{-2\sigma T_e}z^{-2}}$
$\exp(-\sigma t)\sin(\omega_0 t+\varphi_0)$U($t$) with $\sigma > 0$	$y[k] = \exp(-\sigma kT_e)\sin(\omega_0 kT_e+\varphi_0)$ $\forall k \geq 0$	$\frac{\sin(\varphi_0)+e^{-\sigma T_e}z^{-1}\sin(\omega_0 T_e-\varphi_0)}{1-2e^{-\sigma T_e}z^{-1}\cos(\omega_0 T_e)+e^{-2\sigma T_e}z^{-2}}$

Table 1.1. *Continuous-time, sampled signals and their z-transforms*

However, the ZT, such as that presented above, cannot be applied to distributions such as the Dirac impulse $\delta(t)$, as the “Dirac comb” distribution is already present in the definition, the signal would therefore be sampled twice. Such an operation would not make sense and it is thus necessary to substitute $\delta(t)$ with an ordinary function homogeneous to the reciprocal of time and that possesses an area under the curve equal to 1, in order to fulfill the properties of $\delta(t)$. By convention and because there is only a single sample per sampling period, a time step of duration T_e and amplitude $1/T_e$ is chosen, which ensures the consistency of the computations of the ZT of the impulse response of continuous-time systems with those of already sampled systems.

Properties of the ZT defined by $ZT\{y(t)\} = \sum_{k=0}^{\infty} y[k]\, z^{-k}$

– *Lag* (with $x[k]$ causal) : if $y[k] = x[k - k']$ (with $k' > 0$), $Y(z) = z^{-k'} X(z)$

– *Lead*: if $y[k] = x[k + k']$ (with $k' > 0$), $Y(z) = z^{k'} \left[X(z) - \sum_{k=0}^{k'-1} x[k] z^{-k} \right]$

– *Time summation, integration, interpolation*: by applying the definition to $y[k] = \sum_{i=0}^{k} x[i]$, it is shown that $Y(z) = X(z) \frac{1}{1 - z^{-1}}$. As a matter of fact:

$$\mathrm{TZ}\{y(t)\} = \sum_{k=0}^{\infty} y[k]\, z^{-k} = \sum_{k=0}^{\infty} \sum_{i=0}^{k} x[i]\, z^{-k} = \sum_{k=0}^{\infty} \left(\sum_{i=0}^{k} x[i] z^{-i} \right) z^{-(k-i)}$$

$$= x[0] + (x[0] + x[1]) z^{-1} + (x[0] + x[1] + x[2]) z^{-2} + \cdots = x[0] \sum_{k=0}^{\infty} z^{-k} + x[1] \sum_{k=1}^{\infty} z^{-k} + x[2] \sum_{k=2}^{\infty} z^{-k} + \cdots$$

$$= x[0] \sum_{k=0}^{\infty} z^{-k} + x[1] z^{-1} \sum_{k=0}^{\infty} z^{-k} + x[2] z^{-2} \sum_{k=0}^{\infty} z^{-k} + \cdots = \left(x[0] + x[1] z^{-1} + x[2] z^{-2} + \cdots \right) \sum_{k=0}^{\infty} z^{-k} = \frac{X(z)}{1 - z^{-1}}.$$

For a pure integrator (of transmittance $1/s$ obtained by LT in the continuous-time domain), we can calculate an approximation to the integral of the samples using the right rectangle method, and by recurrence: $y[k] = y[k-1] + T_e\, x[k]$. The ZT provides the corresponding transmittance $\frac{Y(z)}{X(z)} = \frac{T_e}{1 - z^{-1}}$. If rather $x[k-1]$ is chosen (left rectangle method), we obtain $\frac{Y(z)}{X(z)} = \frac{T_e z^{-1}}{1 - z^{-1}}$, whereas the trapeze method makes use of the interpolation $y[k] = y[k-1] + T_e\, (x[k] + x[k-1])/2$, whose ZT is: $\frac{Y(z)}{X(z)} = \frac{T_e \left(1 + z^{-1}\right)}{2\left(1 - z^{-1}\right)}$.

Concerning the same integrators which directly manage the samples in the discrete-time domain, the same ZT is obtained but devoid of factor T_e. They are often represented by an operation diagram incorporating a delay of one sampling period and additive looping (Figure 1.45), corresponding to the plus sign appearing in the second member of the preceding recurrence relations.

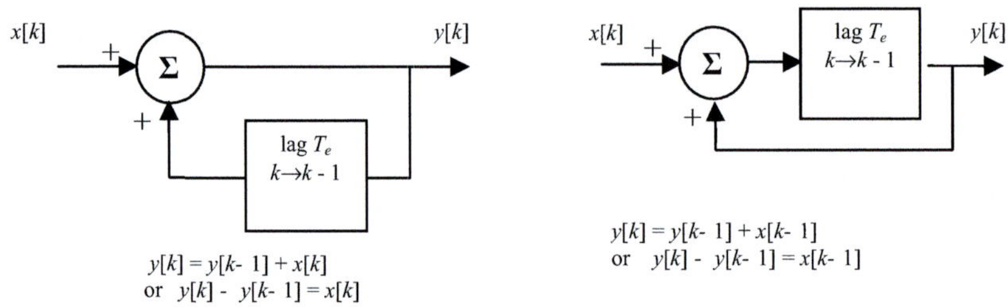

Figure 1.45. *Block diagram of the sampled simple integrators*

More accurate approximations require that the area included under arcs of parabola or polynomial functions of degree >2 be determined, thus implementing interpolation using a number of samples equal to the degree to which 1 is added.

For example, for an arc of parabola, employing the Simpson method, $y[k] = y[k–2] + T_e\,(x[k] + 4\,x[k–1] + x[k–2])/3$, whose ZT yields $\dfrac{Y(z)}{X(z)} = \dfrac{T_e\left(1+4z^{-1}+z^{-2}\right)}{3\left(1-z^{-2}\right)}$.

– *Derivation and difference between successive samples*, namely $y[k] = x[k] - x[k–1]$, in the discrete-time domain. The ZT is simply $Y(z) = X(z)\,(1 - z^{-1})$, the inverse operation of that which applied to the summation in the discrete-time domain. Since neither the derivation $\dfrac{dx(t)}{dt} = \lim\limits_{\Delta t \to 0}\left[\dfrac{x(t+\Delta t)-x(t)}{\Delta t}\right]$ nor the derivative of the Dirac impulse applicable to distributions (see Appendix of Volume 2 [MUR 17b]) can operate on a sequence of samples, because the difference calculated between a sample and itself is zero, resulting from $\Delta t \to 0$ in the derivative definition, Δt is replaced by T_e. The derivation operator then becomes $(1 - z^{-1})/T_e$. More generally, nth-order derivatives in the continuous-time domain are replaced by the differences $\dfrac{x[k]-x[k-n]}{T_e^n}$ in the discrete-time domain, whose ZT are $\dfrac{X(z)-z^{-n}X(z)}{T_e^n}$.

– *Derivation in the complex z-plane of the ZT*: by deriving z^{-k} we obtain $-k\,z^{-k-1}$ from which $\dfrac{dX(z)}{dz}=\dfrac{d\left(TZ\{x(t)\}\right)}{dz}=Y(z)=-\sum_{k=0}^{\infty}k\,x[k]\,z^{-k-1}=\sum_{k'=0}^{\infty}y[k']\,z^{-k'}$ which corresponds by identifying the coefficients of the powers of z to $k' = k+1$, and to $y[k+1] = -k\,x[k]$.

– *Poles*: for the first-order transmittance (Table 1.1), the denominator cancels for $z = z_1 = \exp(-\alpha T_e)$, corresponding to a real pole $-\alpha$ in the plane of the s variable. The roots of the denominator of the last four second-order expressions of the previous table multiplied by $e^{2\sigma T_e}z^2$ in order to obtain the poles directly in the z-plane are $z_{\pm}=e^{-\sigma T_e}e^{\pm j\omega_0 T_e}$, which are complex conjugates and correspond to the poles $-\sigma \pm j\omega_0$ in the plane of the s variable. In these denominators, we recognize the opposite of the sum of poles as the coefficient of z^{-1} and the product of the poles in the z-plane as the coefficient of z^{-2}.

– *Transmittance stability criterion in rational fractions F(z)*: it follows from the Nyquist criterion that indicates that there should be no poles of $F(s)$ in the right half-plane of the complex plane of the s variable, limited by $s = j\omega$, when ω is going from $-\infty$ to $+\infty$. Applying the transformation $z = \exp(j\omega T_e)$, the image of z in the complex plane of the z variable follows a circle of radius 1 when ω goes from $-\pi f_e$ to $+\pi f_e$, then repeatedly beyond this interval. The poles $p_i = \sigma_i \pm j\omega_i$ with positive real part σ_i give poles $z_i = \exp(\sigma_i T_e)\exp(\pm j\omega_i T_e)$ in the z-plane, and their images are located outside the circle of radius 1, which cause instability of the system. It should be noted that on the contrary $p_i = -\sigma_s \pm j\omega_s$ with $\sigma_s > 0$ involves poles with negative real parts in the plane of the s variable, characteristic of a stable system and gives poles inside the circle of radius 1 in the z-plane:

A system is stable if its rational transmittance F(z) has only poles inside the circle of radius 1 in the complex plane of the z variable.

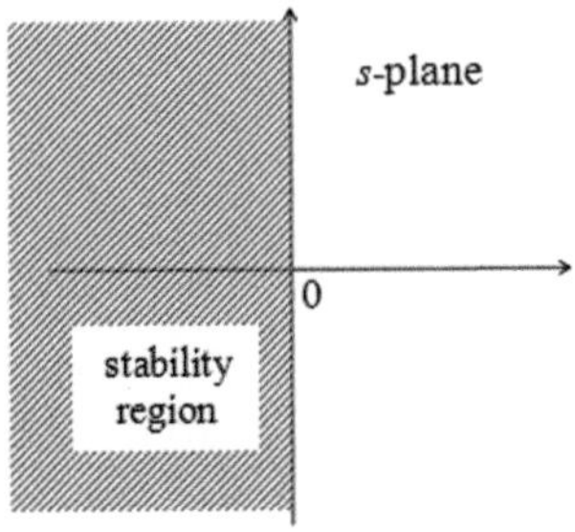

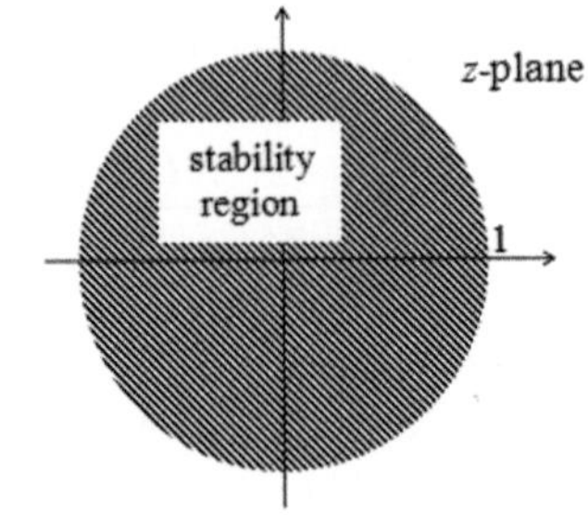

The elementary first-order transmittance $\dfrac{1}{1-e^{-\alpha T_e}z^{-1}}$ is thus stable if $\alpha > 0$, and since the modulus of the poles of the second-order elementary transmittances is $e^{-\sigma_s T_e}$, these transmittances are equally stable with $\sigma_s > 0$, which corresponds to a damped time response either converging towards 0 or towards a constant, that is also non-divergent.

– *Initial and final values*: when $z \to +\infty$, all terms having non-zero powers of z^{-1} cancel out for a convergent series $Y[z]$ and therefore only the first term remains, that is:

$$\lim_{z\to+\infty} Y(z) = y[0], \quad \text{initial value of the time series.}$$

According to the lead theorem: TZ$\{y(t+T_e) - y(t)\} = z\,(Y(z) - y[0]) - Y(z)$

but also TZ$\{y(t+T_e) - y(t)\} = \sum_{k=0}^{\infty}\left(y[k+1]-y[k]\right)z^{-k}$ which tends to $y[\infty] - y[0]$ when $z \to 1$, which delivers the final value of the series: $\lim_{k\to+\infty} y[k] = \lim_{z\to 1}(z-1)Y(z) = \lim_{z\to 1}(1-z^{-1})Y(z)$ from the comparison of both results.

– *Inverse transform*: if we limit ourselves to the values of the original function evaluated only at the sampling times, two techniques can be employed, the second being particularly direct:

- Partial fraction decomposition of the rational function and identification with items found in the dictionary of ZTs (Table 1.1).

- Division of the numerator by the denominator according to the increasing powers of z^{-1} from 0, that is to say according to the increasingly negative powers of z. According to the definition of ZT, the sample sequence then simply consists of the coefficients of $z^0, z^{-1}, z^{-2}, z^{-3}, \ldots z^{-k}$.

1.4.2. *Continuous-time systems subject to a sampled signal*

For a causal continuous-time system whose impulse response is $h(t)$, the response $y(t)$ to an input $x(t)$ is given by the convolution product (Chapter 2 of Volume 2 [MUR 17b]) $y(t)=\int_0^{\infty} h(\tau)x(t-\tau)\,d\tau$. If definition $Y(z)=\int_0^{\infty}\sum_{k=-\infty}^{\infty}\delta(t-kT_e)\,y(t)\exp(-pt)\,dt$ is applied to this convolution product in order to determine the ZT of the sampled response, it still follows that $Y(z)=\sum_{k=0}^{\infty} y[k]\,z^{-k}$.

This expression depends on the discrete samples $y[k]$, which are therefore to be determined. In the same way that linearity and stationarity properties allow us to calculate the output of a continuous-time system by the convolution of the input signal and the impulse response (see Chapter 2 of Volume 2 [MUR 17b]), they can be defined by the discrete convolution:

$y[k]=\sum_{i=0}^{\infty} x[i]h[k-i]=\sum_{i=0}^{\infty} h[i]x[k-i]$. Then TZ$\{y(t)\}$ =

$$\sum_{k=0}^{\infty}\sum_{i=0}^{\infty} x[i]h[k-i]\;z^{-k}=\sum_{k=0}^{\infty}\sum_{i=0}^{\infty} x[i]\;z^{-i}\;h[k-i]\;z^{-(k-i)}=\sum_{i=0}^{\infty} x[i]\;z^{-i}\sum_{k=0}^{\infty} h[k-i]\;z^{-(k-i)}$$

$=\sum_{i=0}^{\infty} x[i]\;z^{-i}\sum_{k=0}^{\infty} h[k]\;z^{-k}=X(z)Y(z)$ because the samples $h[k-i]$ are 0 for $k<i$ for a causal system.

Conclusion: the product of the z-transmittance of either a continuous or sampled system by the ZT of the input signal thus provides the ZT of the output signal.

The z-transmittance of any transfer function involving any physical quantity can thus be determined. In the case where the corresponding impulse response $h(t)$ is available, we apply the definition of the ZT given at the beginning of section 1.4.1.

When we know the Laplace transmittance $T(s) = \mathrm{TL}\{h(t)\}$, the simplest method is to return to the corresponding impulse response (see tables of Chapters 1 and 2, Volume 2 [MUR 17b]) and to sample it, or more precisely to calculate $h(kT_e)$, which is normalized by the period T_e, to obtain $h[k]$ then $\sum_{k=0}^{\infty} h[k]\, z^{-k}$. In effect, the convolution in the continuous-time domain is an integral whose result must be expressed with the same units as the input signal when the system does not change the dimension of the input quantity, which requires that $h(t)$ be inverse time-homogeneous in this case. Nonetheless, this no longer holds with the discrete convolution and $h[k]$ should be calculated as an integral of $h(t)$ over the period T_e. In fact, we will always use the rectangle approximation by simply carrying out the product of $h(kT_e)$ by T_e. This is consistent with the respective dimensions of $\delta(t)$ (inverse-time) and $\mathsf{U}(t)$ (dimensionless) that appear in $h(t)$. The case of analog filters is treated in the following table based on the impulse responses given in Chapter 2 of Volume 2 [MUR 17b] and the ZT dictionary of section 1.4.1.

$H(s)$	Impulse response $h(kT_e)$	$H(z)$
$\dfrac{\omega_c}{s+\omega_c}$ (first order low-pass)	$\omega_c \mathsf{U}(t) \exp(-\omega_c kT_e)$	$\dfrac{\omega_c T_e}{1-e^{-\omega_c T_e} z^{-1}}$
$\dfrac{s}{s+\omega_c}$ (first order high-pass)	$\delta(t) - \omega_c \mathsf{U}(t) \exp(-\omega_c kT_e)$	$\dfrac{1-\omega_c T_e - e^{-\omega_c T_e} z^{-1}}{1-e^{-\omega_c T_e} z^{-1}}$
$\dfrac{\omega_n^2}{s^2+2\zeta\omega_n s+\omega_n^2}$ (second order low-pass)	$\mathsf{U}(t)\dfrac{\omega_n \exp(-\zeta\omega_n kT_e)}{\sqrt{1-\zeta^2}}\sin\left[\omega_n\sqrt{1-\zeta^2}\,kT_e\right]$	$\dfrac{\left(\dfrac{\omega_n T_e}{\sqrt{1-\zeta^2}}\right)e^{-\zeta\omega_n T_e}\sin\left[\omega_n\sqrt{1-\zeta^2}\,T_e\right]z^{-1}}{1-2e^{-\zeta\omega_n T_e}\cos\left[\omega_n\sqrt{1-\zeta^2}\,T_e\right]z^{-1}+e^{-2\zeta\omega_n T_e}z^{-2}}$
$\dfrac{s\omega_n}{s^2+2\zeta\omega_n s+\omega_n^2}$ (second order band-pass)	$\mathsf{U}(t)\dfrac{\omega_n \exp(-\zeta\omega_n kT_e)}{\sqrt{1-\zeta^2}}\cos\left[\omega_n\sqrt{1-\zeta^2}\,kT_e+\varphi_0\right]$ with $\sin(\varphi_0)=\zeta$	$\dfrac{\left(\dfrac{\omega_n T_e}{\sqrt{1-\zeta^2}}\right)\left[\cos(\varphi_0)-z^{-1}e^{-\zeta\omega_n T_e}\cos(\omega_n\sqrt{1-\zeta^2}\,T_e-\varphi_0)\right]}{1-2e^{-\zeta\omega_n T_e}\cos\left[\omega_n\sqrt{1-\zeta^2}\,T_e\right]z^{-1}+e^{-2\zeta\omega_n T_e}z^{-2}}$
$\dfrac{s^2}{s^2+2\zeta\omega_n s+\omega_n^2}$ (second order high-pass)	$\delta(t) -$ $\mathsf{U}(t)\dfrac{\omega_n \exp(-\zeta\omega_n kT_e)}{\sqrt{1-\zeta^2}}\sin\left[\omega_n\sqrt{1-\zeta^2}\,kT_e+2\varphi_0\right]$	$1 -$ $\dfrac{\left(\dfrac{\omega_n T_e}{\sqrt{1-\zeta^2}}\right)\left[\sin(2\varphi_0)+z^{-1}e^{-\zeta\omega_n T_e}\sin(\omega_n\sqrt{1-\zeta^2}\,T_e-2\varphi_0)\right]}{1-2e^{-\zeta\omega_n T_e}\cos\left[\omega_n\sqrt{1-\zeta^2}\,T_e\right]z^{-1}+e^{-2\zeta\omega_n T_e}z^{-2}}$

Table 1.2. *Impulse response and z-transform of elementary analog filters*

1.4.3. *Switched-capacitor circuits and infinite impulse response (IIR) filters*

1.4.3.1. *Analysis of the effects of switching on the charges stored in capacitors and the passive first order low-pass filter*

MOS switches associated with capacitors and eventually operational amplifiers provide means to build all types of z-transmittance. Type P and I switches (MOS transistors) are respectively switched off during the half sampling period when the control is 0 and switched on (closed contact) when it is 1. P is in phase with integer times and I in inverse phase, that is, switched on in synchronization with half-integer times (see Figure 1.46). There is no overlap of closures over time ("break-before-make switch") and charge transfers are supposed to be immediate because the resistances of the switch are neglected.

Example of the passive low-pass filter:

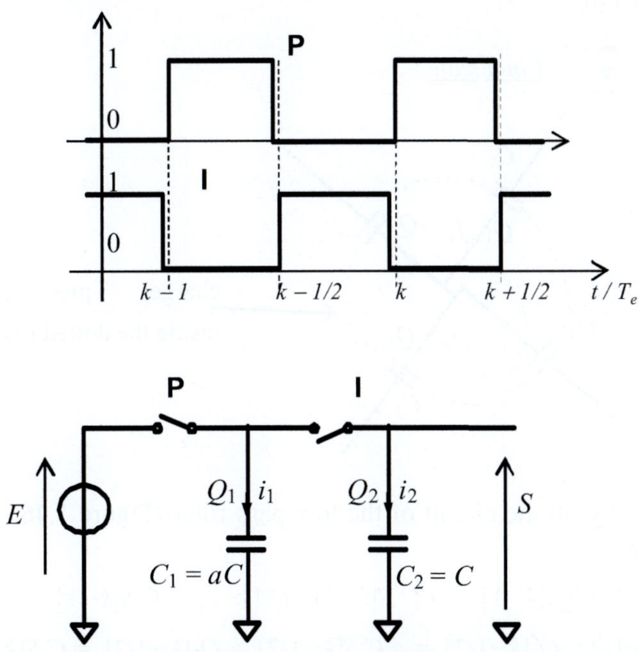

Figure 1.46. *Switch status (0 = switched off or open; 1 = switched on or closed) and circuit of the low-pass filter*

These circuits must be analyzed based on two types of relations that, on the one hand, can be applied in early P phases and I phases, and, on the other hand, during the two transitions from one phase to the other:

– Relation between the charge Q and the difference of potential V at the terminals of each capacitor of capacitance C: $Q = CV$.

– Conservation of the overall charge in a system isolated during switching; by integrating the Kirchhoff law of current conservation, we get zero change for the total charge, for example, in the case of four capacitors connected to a node:

$$\int_{(k-1)T_e}^{(k-1/2)T_e} (i_1 + i_2 + i_3 + i_4)dt = 0 = \left[Q_1 + Q_2 + Q_3 + Q_4\right]_{(k-1)T_e}^{(k-1/2)T_e}$$

and $$\int_{(k-1/2)T_e}^{kT_e} (i_1 + i_2 + i_3 + i_4)dt = 0 = \left[Q_1 + Q_2 + Q_3 + Q_4\right]_{(k-1/2)T_e}^{kT_e}$$

Final state :

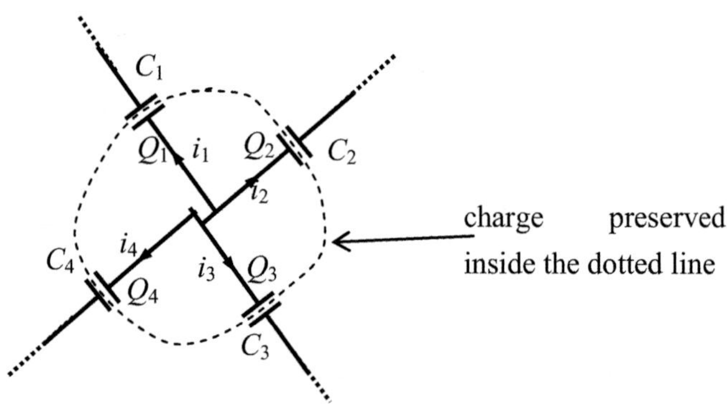

Consequently, in the circuit of the low-pass filter (Figure 1.46), we have the following:

– time $k-1$: $Q_1[k-1] = a\,C\,E[k-1]$; $Q_2[k-1] = C\,S[k-1]$;

– time $k-1/2$: $Q_1[k-1/2] = a\,C\,S[k-1/2]$; $Q_2[k-1/2] = C\,S[k-1/2]$;

and charge conservation leads to:

$$\int_{(k-1)T_e}^{(k-1/2)T_e} (i_1 + i_2)dt = 0 = [Q_1 + Q_2]_{(k-1)T_e}^{(k-1/2)T_e} = Q_1[k-1/2] + Q_2[k-1/2] - Q_1[k-1] - Q_2[k-1];$$

– time k: $S[k] = S[k-1/2]$ because C_2 preserves its charge when I opens.

Hence the recurrence relation: $(1 + a)\, C\, S[k] - C\, S[k-1] = a\, C\, E[k-1]$.

Since the terms of this equation are products of capacitance by voltage, this equation expresses charge conservation.

The ZT is simply $(1 + a)\, S(z) - z^{-1}\, S(z) = a\, z^{-1}\, E(z)$ simplifying by C.

Hence the z-transmittance: $T(z) = \dfrac{S(z)}{E(z)} = \dfrac{a}{1+a} \dfrac{z^{-1}}{1 - \dfrac{1}{1+a} z^{-1}}$.

For sinusoidal signals $z = \exp(j\omega T_e)$ and if $\omega T_e << 1$, it can be written at the first order that $z^{-1} = 1 - j\omega T_e$. Hence:

$$T(j\omega) \approx \frac{a}{1+a} \frac{1 - j\omega T_e}{1 - \dfrac{1}{1+a}(1 - j\omega T_e)} \approx \frac{1}{a} \frac{1}{1 + j\omega T_e \dfrac{1+a}{a}}.$$

This represents a first-order low-pass transmittance with time constant $T_e \dfrac{1+a}{a}$, which corresponds to arranging the two capacitors $C + aC$ in parallel for the capacitive factor. The resistive factor is thus equal to $\dfrac{T_e}{aC}$, or more precisely a resistance proportional to the sampling period. The step response can be obtained by:

$$S_U(z) = T(z) \frac{1}{1 - z^{-1}} = \frac{abz^{-1}}{\left(1 - bz^{-1}\right)\left(1 - z^{-1}\right)} = \frac{1}{1 - z^{-1}} - \frac{1}{1 - bz^{-1}} \text{ defining } b = \frac{1}{1+a} < 1$$

after partial fraction expansion.

The first term corresponds to the ZT of the Heaviside step function and the second term to the ZT of successive samples of amplitude b^k according to the table of transforms. It is also possible to find this result by carrying out the division of the numerator by the denominator expanded from the fraction not decomposed, that is $S_U(z)=\sum_{k=1}^{\infty}\left(1-b^k\right)z^{-k}$.

The amplitudes of the samples $s_U[k] = 1-b^k$ are the coefficients of z^0, z^{-1}, $z^{-2}, z^{-3}, \ldots z^{-k}$:

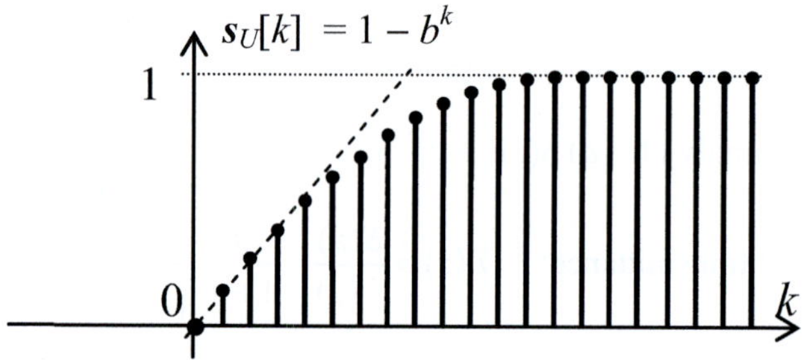

Figure 1.47. *Step response of the first-order low-pass filter*

The straight line passing through the first two samples (Figure 1.47), equivalent to the tangent at the origin in the continuous-time domain, cuts the ordinate 1 in $k \approx \dfrac{1}{1-b} = \dfrac{1+a}{a}$, which again correctly yields a time constant $T_e\dfrac{1+a}{a}$ if we make the analogy with the response of the continuous-time first-order low-pass circuit.

1.4.3.2. *Assembly using operational amplifiers and charge transfer transmittances; basic IIR (infinite impulse response) filters*

The use of an operational amplifier (or several) and negative feedback magnifies the possibilities. The basic circuit for obtaining the first-order transmittance, of the first degree in z in the numerator and in the denominator, can be found in Figure 1.48.

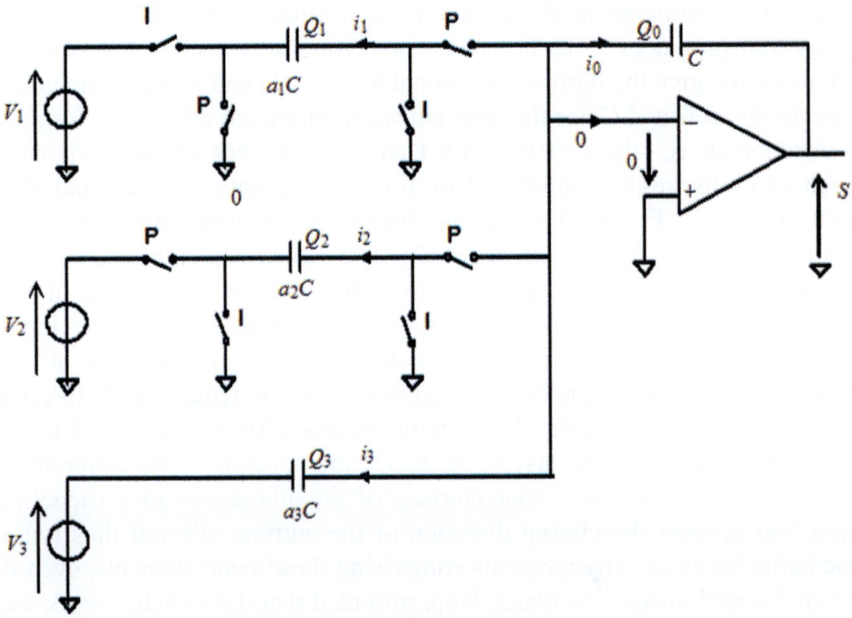

Figure 1.48. *First-order universal filter*

We calculate Figure 1.48 with the same method as previously seen (see exercise 1.6.2) and based on the assumption of an ideal operational amplifier:

$$C\ S[k] = C\ S[k-1] + a_1\ C\ V_1[k-1] - a_2\ C\ V_2[k] - a_3\ C\,(V_3[k] - V_3[k-1])$$

By the ZT: $C\,S(z) = z^{-1}\,C\,S(z) + a_1\,C\,z^{-1}\,V_1(z) - a_2\,C\,V_2(z) - a_3\,C\,(1 - z^{-1})\,V_3(z)$

or still $0 = a_3\,C\,(z^{-1} - 1)\,V_3(z) + C\,(z^{-1} - 1)\,S(z) + a_1\,C\,z^{-1}\,V_1(z) - a_2\,C\,V_2(z)$.

In the final form of this equation, each term of the second member originates from a branch located between a voltage source and the minus input of the operational amplifier and the first member represents the zero current on the operational amplifier input. Taking into account the definition of the ZT, the fact that the variable z is dimensionless, and the coefficients of the voltages that are capacitances, it can be considered that each term is a charge transfer operated by the switches in each capacitor between the voltage generator, on the one hand, and the input of the operational amplifier, on the other, during a sampling period. The voltage coefficients can thus be considered in the equation transformed by the ZT as "charge transfer transmittances". The branches comprising only a single capacitance with no switch give the terms proportional to $(z^{-1} - 1)$ and to the capacitance (respectively $a_3\, C$ and C for the first and second terms of the second form of the equation in z); the system with I and P switches on the horizontal branch gives the term proportional to $a_1\, C\, z^{-1}$ (capacitance $a_1\, C$) and that which uses two P switches in the horizontal branch gives the term proportional to $-\, a_2\, C$ (capacitance $a_2\, C$). The global equation results from the conservation of currents and thus from the charges at the node of the minus input of the operational amplifier with a zero value of the input current. Each term is also proportional to the potential difference at the terminals of the element under consideration, which is reduced here to each input voltage source because the operational amplifier is considered to be ideal. In the first case, it is easy to recognize the opposite of the differential operator $(1 - z^{-1})$, which is characteristic of the admittance of a capacitor taking into account the chosen direction of the current. We can thus reuse these terms for all the arrangements comprising these same elements, even if they are located inside a feedback loop, provided that the switches work the same way as in Figure 1.48.

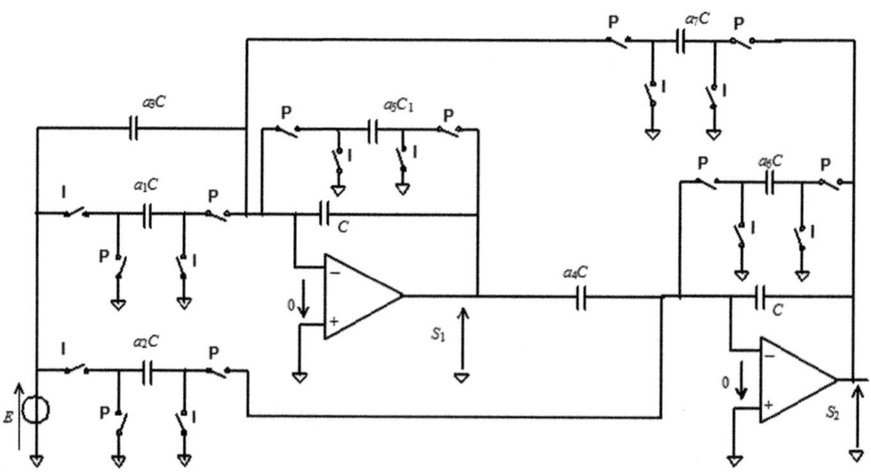

Figure 1.49. *Typical second-order assembly*

To facilitate the notations of the charge conservation equations, we can build a functional diagram constituted by the charge transfer transmittances from all branches and the adders imposing that currents and voltage be equal to 0 on the input of operational amplifiers (see section 1.5 and exercise 1.6.2).

By adding a second assembly based on operational amplifiers and by re-looping on the minus input of the first operational amplifier, we obtain the second-order transmittance, of the second degree in z in the numerator and in the denominator (Figure 1.49 and exercise 1.6.2).

The basic second-order filter thus constructed has a transfer function in z that is written in the following general way:

$$H_i(z)=\frac{b_{1i}+b_{2i}z^{-1}+b_{3i}z^{-2}}{d_{1i}+d_{2i}z^{-1}+d_{3i}z^{-2}}=\frac{(z-z_{ni})(z-\overline{z}_{ni})}{(z-z_{di})(z-\overline{z}_{di})}.$$

Provided that a branch is introduced and located between the source connected to the input and the second operational amplifier (a_2C), as in Figure 1.49, we can choose the relative sign of the coefficients of the numerator and denominator of this transmittance in z (see exercise 1.6.2). By cascading these arrangements, it thus becomes possible to obtain second-, fourth-, sixth-order-; and so on, transmittances with easily computable poles. It is also possible to use a multiple feedback loop-based assembly and to build ladder filters in which each cell, most often of the second order, is equivalent to one among those studied in Chapter 2 of Volume 2 [MUR 17b] for image-matching filters. These cells can incorporate, on the one hand, capacitor admittances proportional to the differential operator $(z^{-1}-1)$ apart from the sign as previously reported, and, on the other hand, inductance admittances proportional to the integration operator $(z^{-1}-1)^{-1}$ that have to be developed with an operational amplifier. Strategies that take into account and fix most imperfections of this type of active component and CMOS switch must be implemented (refer to [BAI 96], because a more detailed study of these systems will not be discussed here). We can thus produce ladder filters between two terminations when the solution to matching problems has to be optimized.

Summary: filters based on two capacitors and switched-capacitor circuits including, in addition, an operational amplifier with a negative feedback loop provide z-transfer functions which are rational fractions of the first or second degree in z^{-1} (see exercises) that can be easily cascaded. The division of the numerator by the denominator leads to unlimited expansion comprising an infinite number of terms, and consequently these filters are *characterized by an infinite impulse response (IIR). They allow for cutoff or central frequencies to be obtained which are adjustable and proportional to the sampling frequency.*

1.4.3.3. *Non-switched electrical circuits subject to sampled signals*

In mixed arrangements there are non-switched analog filters that are subject to sampled signals because of a switched element that is present upstream in the operational chain. This is the case of the PLL. With a few precautions, we can extend the approach of the previous subsection to the branches also containing resistors. For a branch submitted to the voltage $V[k]$ and comprising only a single resistance $R = 1/G$, the charge transfer is done by way of the current $G\ V[k]$.

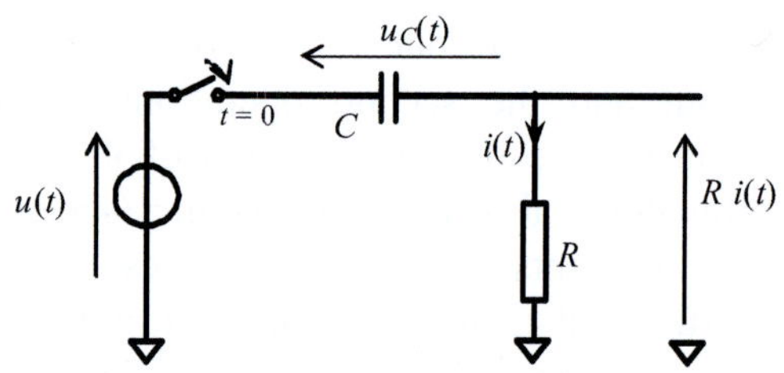

Figure 1.50. *CR circuit*

For a branch including a resistance in series with a capacitor, it is best to restart from an analysis in the time-continuous domain. In both cases, it can be assumed that the result depends on the time the voltage is maintained and/or the current can flow. In the case of a CR branch (capacitance-resistance) as in Figure 1.50, subject to a step of voltage $u(t) = V_0(t)\ \mathrm{U}(t)$, where $\mathrm{U}(t)$ is the unit step, we must solve the equation $i(t) + RC\dfrac{di(t)}{dt} = C\dfrac{du(t)}{dt}$ whose solution is (see Appendix in Volume 2 [MUR 17b]): $i(t) = \dfrac{V_0}{R}\exp\left(-\dfrac{t}{RC}\right)\mathrm{U}(t) - \dfrac{V_1}{R}$, with $CV_1 = q(0^-)$, initial charge at the terminals of the capacitor corresponding to a voltage $u_C(0^-) = V_1 = u(0^-)$ when there is no initial current[1].

1 Note: it is necessary here that the circuit be opened for $t < 0$ in order to preserve the previous charge in the capacitance C which, in the opposite case, would discharge when $u(t) =$

We can also calculate $u_C(t) = u(t) - Ri(t) = V_0\left[1-\exp\left(-\frac{t}{RC}\right)\right]\mathrm{U}(t)+V_1$.

If the voltage $u(t)$ is a very brief impulse compared to RC, equivalent to a sampled signal in the strict sense, there will be no change in the charge in the capacitor due to the cancellation of the first term in the expression of $u_C(t)$. This shows that the effect of the resistance R should be minimized as much as possible with respect to $T_e/2$, which is the closing time of the switches in the switched-capacitor assemblies of previous sections. When, on the contrary, this resistance has a consistent role in defining a time constant useful for filtering, the applied voltages should not be strictly sampled but sampled and held throughout the whole of the sampling period T_e. It is nearly always the case in mixed arrangements and we will therefore only discuss the case where the charging starts at time 0 and stops at time $t = T_e$. The change in the charge ΔQ_{RC} can thus be calculated as:

$$\Delta Q_{RC} = \int_0^{T_e} i(t)dt = \int_0^{T_e}\left[\frac{V_0}{R}\exp\left(-\frac{t}{RC}\right)\mathrm{U}(t)-\frac{V_1}{R}\right]dt = CV_0\left[1-\exp\left(-\frac{T_e}{RC}\right)\right]-\frac{T_e}{R}V_1.$$

Since the system is stationary, the recurrence equation between $\Delta Q[k]$ and the voltages $U[k] = V_0$ and $U[k-1] = V_1$ can then be deduced by defining $\tau = RC$, $b = \exp\left(\frac{T_e}{\tau}\right)$ and $\beta = T_e/\tau$:

$$\Delta Q_{RC}[k] = C\left[\left(1-b^{-1}\right)U[k]-\beta U[k-1]\right]$$

yielding by the ZT: $\Delta Q_{RC}(z) = C\left[\left(1-b^{-1}\right)-\beta z^{-1}\right]U(z)$.

When $\beta << 1$, which is almost always the case to obtain filtering in the frequency domain where the sampling theorem is valid, $1-b^{-1} \approx 1-(1-\beta) = \beta$ at first order, and $\Delta Q_{RC}[k]$ and $\Delta Q_{RC}(z)$ are respectively simplified into $\Delta Q_{RC}[k] \approx C\beta(U[k]-U[k-1])$ and

---

U(t) = 0 is applied for $t < 0$. However, the switch may be removed when the system is subjected to repetitive unit steps of duration T_e (sampled-and-held signal very common in practice), the previous ending at time $t = 0^-$, because the duration of the capacitance discharge then becomes negligible.

$\Delta Q_{RC}(z) \approx C\beta\left(1 - z^{-1}\right)U(z)$, which is identical to what would be obtained with a capacitor of capacitance $\beta\, C$ instead of the capacitance a_3C of the circuit of Figure 1.50.

For a resistance alone, under the same conditions, $\Delta Q = \int_0^{T_e} i(t)dt = \int_0^{T_e} \frac{V_0}{R} dt = \frac{T_e}{R} V_0$, or more specifically $\Delta Q_R[k] = C\beta U[k]$, which gives by the ZT $\Delta Q_R(z) = C\beta U(z)$.

1.4.4. *Switched-capacitor circuits adapted to finite impulse response (FIR) filters*

Finite impulse response (FIR) filters have the advantage of synthesizing directly from the desired impulse response since the latter is a series in which the coefficients of the successive terms in z^{-k} are the amplitudes of the impulse response at times kT_e, with $k \in [0\ ,\ M]$. To implement the filter, it is therefore necessary to create stages delaying the sampled signal of one period T_e and to assign them an attenuation (or eventually amplification) equal to the value of each coefficient of the filter. This delay operator (Figure 1.51) performs for an analog signal, a function equivalent to a master-slave flip-flop for a logic signal and can be based on the following circuit where the switches P and I always follow the same sequence as previously:

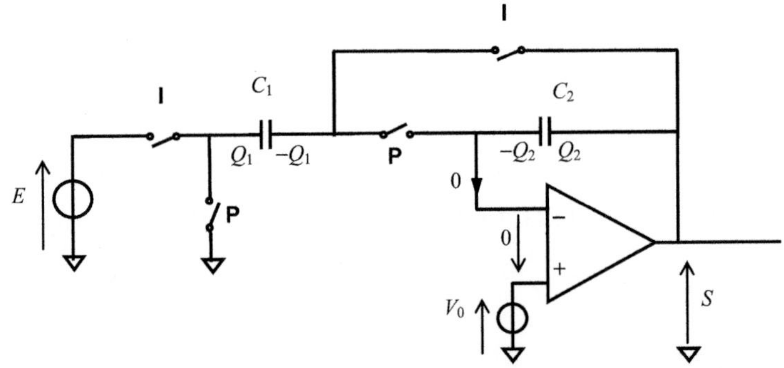

Figure 1.51. *Delay circuit for samples with period T_e*

The offset V_0 of the operational amplifier is taken into account.

– At time $(k-1/2)T_e$, $Q_1[k-1/2] = C_1\,(E[k-1/2] - S[k-1/2])$ and the capacitor C_2 is isolated because of the zero input current of the operational amplifier; therefore:

$$Q_2[k-1/2] = Q_2[k-1] \text{ and } S[k-1/2] = S[k-1].$$

– Considering only the voltages at times kT_e (k integer) and assuming that they do not vary for a period T_e, $E[k-1] = E[k-1/2]$, it follows that:

$$Q_1\,[k-1/2] = C_1\,(E[k-1] - S[k-1]\,).$$

– At time $k\,T_e$, C_1 is charged by the offset voltage V_0 or $Q_1[k\,] = -\,C_1\,V_0$ and $Q_2[k\,] = C_2\,(\,S[k\,] - V_0)$. The armatures of C_1 and C_2 connected by the switch P are isolated; therefore $Q_1[k] - Q_1[k-1/2] + Q_2[k] - Q_2[k-1/2] = 0$.

– Using the previous equations, it follows that:

$$C_1V_0 - C_1\,(E[k-1] - S[k-1]\,) + C_2\,(\,S[k\,] - V_0) - C_2(S[k-1] - V_0) = 0$$

$$C_2\,(\,S[k\,] - S[k-1]\,) + C_1\,S[k-1]\,) = C_1\,(\,E[k-1] + V_0\,).$$

By establishing $a = \dfrac{C_1}{C_2}$ and after ZT:

$$S(z) = a\,\frac{z^{-1}}{1-(1-a)z^{-1}}E(z) + \frac{V_0}{\left(1-(1-a)z^{-1}\right)\left(1-z^{-1}\right)}.$$

If we take $C_1 = C_2$, $a = 1$, then we get: $S(z) = z^{-1}\,E(z) + \dfrac{V_0}{\left(1-z^{-1}\right)}$.

There is thus indeed a delay of one period T_e of the input signal but the offset voltage V_0 is found on output in the form of a step. As it is necessary to arrange M stages of this type in cascade, the offsets will add up from one stage to another and the total could become an annoyance if M exceeds about 10. This assembly is usable provided that it is modified in order to correct the offset (see exercise 1.6.3). To achieve the full FIR filter, it is sufficient to add to the chain of delay cells an adder that weighs every sample delayed by the appropriate coefficient of the impulse response of the filter. In CMOS technology, capacitive dividers and operational amplifiers will be used.

1.4.5. *Sampled systems modeling using functional blocks*

In electronics, a lot of systems deal with electric quantities (currents and voltages). The various ways to determine the *z*-transmittance were detailed in the preceding sections. In the case of sampled systems, the first operation to perform is precisely that of sampling, using a simple sampler (section 1.2.1) or a zeroth-order sample and hold circuit (section 1.2.2). In both cases, the impulse response being a step signal of amplitude 1 in the interval $[0\ \alpha T_e[$ (with $0<\alpha<1$), the ZT is equal to 1 because the impulse response is sampled at the origin and for every period. This result illustrates the fact that *z*-transmittances act on signals already sampled and that renewing this operation brings no further change. Higher order interpolators-samplers achieve the smoothing of signals sampled in the continuous-time domain (section 1.2.3) but their sampled impulse response, which determines their *z*-transmittance in the discrete-time domain, contains as many non-zero samples as their order. These samples are the coefficients of the numerator of the *z*-transmittance, while its denominator is equal to 1 since there is no operation on the output samples (non-recursive filter). As will be seen later, for this reason they fall under the category of FIR filters or even of moving average (MA) filters, but with a very insufficient number of coefficients to have a comparable action with that of FIR filters in general. They will therefore not be considered here. In order to get more efficient smoothing in the discrete-time domain, it will be necessary to implement recursive filters (IIR) that perform one or more integrations (see Figure 1.52).

To include all cases, it is necessary to also consider sampled systems regardless of the dimension of the signals, which are not necessarily electric currents or voltages. For instance, this is the case of the PLL for which the block diagram deduced from a continuous-time approximation transformed by the LT acts on phases and frequencies. Insofar as these quantities are actually measured only once per period, it is legitimate to consider such a system to be sampled, provided that the requirement that the sampling theorem be followed is enforced. This is achieved when the phase or frequency modulation is significantly slower than the frequency of the electrical signals but can be irrelevant for complex signals passing through the same amplitude several times during the fundamental period. These signals then have a spectrum that extends beyond the Nyquist frequency, as is often the case for digital modulations, exhibiting a very wide range around the carrier frequency, and they can only be taken into account and validly handled by the PLL if they have undergone a frequency filtering beforehand intended to limit their spectral range. This condition is none other than the general precondition necessary for a signal to be processed by a sampled system. There is also another solution where the sampling frequency is forcibly maintained at a constant value by a clock which is both much faster than the signals received by the PLL and independent of them, without fundamentally changing the circuits utilized for the detection of the phase shift. In this latter case, or under the assumption that the sampling period can be regarded as being constant, a sampled model of the PLL can be developed and a transmittance in the domain of the z-variable determined (see exercise 1.6.5).

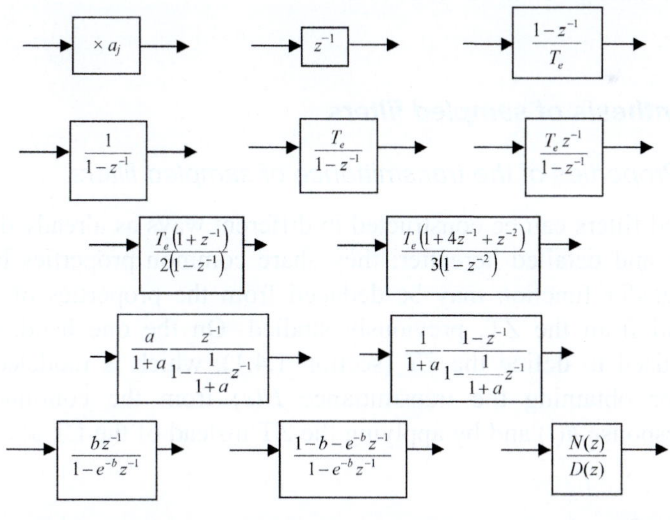

Figure 1.52. *Functional blocks in the domain of the z variable, from left to right and from top to bottom: multiplication by a factor, delay of one sampling period T_e, derivation, summation of successive samples, integration using the right rectangle method, integration using the left rectangle method, integration using the trapezoid method, integration using parabolic interpolation, switched-capacitor-based first order passive low-pass filter (the first capacitance is a times larger than the second), switched-capacitor-based first-order passive high-pass filter, first-order low-pass analog filter, first order analog high-pass filter, with $b = \omega_c T_e$ (where ω_c is the cutoff angular frequency) and all analog filters, digital or switched-capacitor-based whose transmittance in z is known. In order to shift from the derivator to the differentiator, and to obtain the transmittance of the integrators that directly manage samples we omit T_e*

If it is necessary to take into account the variations of the sampling period resulting from frequency modulation for example, this is possible in the context of the state space representation (see section 1.5) provided that the matrix elements be varied according to T_e for every period. This is also the case for nonlinear systems. On the other hand, when the period T_e can be considered as a constant, eventually at the cost of an approximation, a diagram based on functional blocks can be built for any system, each of them being characterized by a z-transmittance and therefore performing a basic operation on the samples, whether they are digital or analog. As a matter of fact, the ZT of a convolution relation between the impulse response and the input signal can be simply obtained by the product of the ZTs of this impulse response, which is called the z-transmittance, and from that of the input signal, as demonstrated in section 1.4.2. The most common cases are given in Figure 1.52 and for second-order analog filters, the reader should refer to the table given in section 1.4.2.

1.4.6. *Synthesis of sampled filters*

1.4.6.1. *Properties of the transmittance of sampled filters*

Sampled filters can be constructed in different ways as already described previously and detailed hereafter; they share common properties however. Their z-transfer function may be deduced from the properties of sampled signals and from the ZT, previously studied. On the one hand, the first approach used to define the ZT (section 1.4.1), which is modeled on the method for obtaining the transmittance $H(s)$ from the continuous-time impulse response $h(t)$ and by applying the ZT instead of the LT allows us to write:

$$H(z) = \int_0^{\infty} \sum_{k=-\infty}^{\infty} \delta(t - kT_e) h(t) \exp(-st)\, dt = \sum_{k=0}^{\infty} h(kT_e) \exp(-skT_e) = \sum_{k=0}^{\infty} h[k]\, z^{-k},$$

where we denote by $h[k]$ the successive samples of the impulse response for k ranging from 0 to infinity. The result is thus the sum of a series of decreasing (and negative) powers of z with $h[k]$ as coefficients. In the case of a FIR filter, there is a finite number of non-zero coefficients allowing for the weighting of the samples delayed by kT_e and it is therefore sufficient to employ them as a multiplicative factor for each sample and then to carry out the summation. However, on the other hand, the impulse response being a sampled signal, its Fourier transform, and thereby its spectrum, is periodic in the frequency domain (see section 1.1.1.4). As a consequence, the transmittance of a sampled filter is periodic in the frequency domain with a frequency equal to the sampling frequency f_e.

According to the real nature of the impulse response and taking into account the properties of symmetry of the transform deriving thereof, we know that the modulus of the transmittance has even symmetry with respect to the zero frequency and that the argument has odd symmetry. Any information related to the transmittance is therefore contained in the interval $[-f_e/2 \;\; f_e/2]$, which is a frequency interval in which the filter is operational according to the sampling theorem. The frequency $f_e/2$ is often referred to as the Nyquist frequency and is used as a normalization frequency. The validity of the sampling theorem implies that the bandwidth of the filter must be contained in this interval, or equivalently, that the sampling frequency be chosen to satisfy this condition and also to exceed the maximal frequency of the significant components contained in the spectrum of the signal to be filtered. The first operation necessary for the synthesis of a filter is therefore the choice of the sampling frequency f_e relative to the characteristic frequency f_c desired for the filter (cutoff frequency or central frequency for bandpass and band-stop filters).

1.4.6.2. *Approximations for shifting from the jω variable to the z variable*

The second synthesis operation involves defining the operational properties of the filter in the form of its transmittance $H(z)$. One of the ways to proceed involves transposing a transmittance $H(s)$ known in the continuous-time domain (see Chapter 2 Volume 2 [MUR 17b]) into the complex plane of the z variable. The first form given for $H(z)$ is that of a series, from which it is not always easy to recover a fraction. In order to obtain a more compact form of $H(z)$ in the form of a rational fraction, a second approach starts from the difference equation governing the linear and stationary system, receiving the samples $x[i+k]$ at input, and delivering samples $y[i+k]$ at output, i being the number of the first sample and k ranging from 0 to n, where n is the order of the filter.

If $\sum_{k=0}^{n} a_{n+1-k}\, y[i+n-k] = \sum_{k=0}^{n} b_{n+1-k}\, x[i+n-k]$, the ZT gives

$H(z) = \dfrac{\sum_{k=0}^{n} b_{n+1-k}\, z^{-n+k}}{\sum_{k=0}^{n} a_{n+1-k}\, z^{-n+k}}$, which can be converted into a product of second-degree fractions (and of one first-degree fraction if n is odd) once the poles and zeros of this fraction are calculated. This form is particularly interesting in the case of IIR filters to make the analogy with a cascade of second-order and first-order filters (or sections) whose z-transmittance have been previously established.

To draw the frequency response $H(j\omega) = H(j2\pi f)$ of the transmittance $H(z)$, it suffices to change variable $z = \exp(j2\pi f T_e) = \exp(j2\pi f / f_e)$ in $H(z)$. Nevertheless, the resulting expression is not a quotient of two polynomials of the z-variable, which prevents the implementation of the synthesis of $H(z)$ with the basic z-transmittances determined previously, since $j2\pi f = j\omega = (1/T_e)\ln(z)$. It is then preferable to proceed by approximation, to be chosen from two possible ones: either the bilinear approximation or the approximation of the preserved sampled impulse response. Each one makes it possible to obtain $H(z)$ as a product of fractions where numerators and denominators are first and second degree polynomials of the z variable.

Bilinear transformation

Bilinear approximation consists of using the rational fraction $\dfrac{z-1}{z+1}$ instead of $\ln(z)$, whose limited expansion is close enough. The geometric transformation of the circle described by the image of the variable $z = \exp(j\omega T_e)$ in the complex plane by means of the expression $\dfrac{z-1}{z+1}$ provides a straight line falling over the imaginary axis, like the image of $j\omega$, as shown in the following calculation. In the basic version, $j\omega$, or more precisely the variable of the transmittance $H(s)$ in harmonic regime, is replaced by $2f_e\dfrac{z-1}{z+1}$, relying on the following calculation. Since

$$\frac{z-1}{z+1}=\frac{\exp(j\omega T_e)-1}{\exp(j\omega T_e)+1}=\frac{\exp(j\omega T_e/2)-\exp(-j\omega T_e/2)}{\exp(j\omega T_e/2)+\exp(-j\omega T_e/2)}=\frac{j\sin(\omega T_e/2)}{\cos(\omega T_e/2)}=j\tan(\omega T_e/2),$$

which is a purely imaginary result, this means that ω is replaced by $2f_e\tan(\omega T_e/2) = \frac{\tan(\omega T_e/2)}{T_e/2}$. Although the image is actually positioned on the imaginary axis, a systematic error results thereof in the frequency and angular frequency, which is progressively augmented as the ratio $\frac{\tan(\omega T_e/2)}{\omega T_e/2}$ increases. For $\omega T_e = \pi/4$, corresponding to a frequency half the Nyquist frequency $f_e/2$, the relative deviation is 5.5 % and goes to 27 % for the Nyquist frequency. This frequency shift phenomenon, which is well known in systems for demodulating single-sideband signals, is colloquially likened to a chirp or to a warp. It is more interesting to reduce it to 0 in the neighborhood of the characteristic frequency of the filter $f_c = \omega_c/2\pi$ by adopting a modified bilinear transformation in which ω is replaced by $\frac{\omega_c}{\tan\left(\omega_c T_e/2\right)}\tan(\omega T_e/2)$, which corresponds to replacing the variable s of the transmittance $H(s)$ by $\omega_{bl}\frac{z-1}{z+1}$ where $\omega_{bl}=\frac{2\pi f_c}{\tan\left(\pi\frac{f_c}{f_e}\right)}$. We verify that for $\omega = \omega_c = 2\pi f_c$, the result of the transformation is recovered without error. The relative error increases on both sides of this frequency f_c, but it is smaller than in the case of the basic transformation, and it does not greatly deteriorate the frequency response, because it either affects the bandwidth in which $H(j\omega)$ is close to a constant or affects the stopband in which $|H(j\omega)|$ is much lower.

It should be noted that this transformation does neither change the degree of the numerator nor change the degree of the denominator, thereby the order of the filter is not changed.

Approximation of the preserved sampled impulse response

When we have the transmittance in the form of a rational fraction, it is possible to decompose it into partial fractions, namely in the form of a sum of fractions of the first and second degree. The corresponding responses in any domain whatsoever are thus the sum of the responses of the basic systems of the first and second order studied in this chapter and those of the previous volumes [MUR 17a, MUR 17b]. It is then easy to make the analogy between continuous-time and sampled impulse responses, by comparing the last column of the table in section 2.2.1 of Chapter 2 of Volume 2 [MUR 17b] and the central column of Table 1.1. For the first order, the identification indicates $\omega_1 = \alpha$, while for the second order, $\zeta\omega_n = \sigma$, $\omega_n\sqrt{1-\zeta^2} = \omega_0$ and $\zeta = \sin\varphi_0$. However, the simple second-order elements are first-degree rational fractions in the numerator and of the second-degree in the denominator, obtained by grouping the complex conjugate poles, which corresponds to a weighted sum of the respective impulse responses of the basic low-pass and band-pass filters (see table in section 2.2.1, Chapter 2 of Volume 2 [MUR 17b]) $h_1(t) = \dfrac{\omega_n \exp(-\zeta\omega_n t)}{\sqrt{1-\zeta^2}} \sin\left(\omega_n\sqrt{1-\zeta^2}\,t\right)$ U(t) and $h_2(t) = \dfrac{\omega_n \exp(-\zeta\omega_n t)}{\sqrt{1-\zeta^2}} \cos\left(\omega_n\sqrt{1-\zeta^2}\,t + \arcsin\zeta\right)$ U(t), corresponding, respectively, to transmittances $\dfrac{\omega_n^2}{s^2+2\zeta\omega_n s+\omega_n^2}$ and $\dfrac{\omega_n s}{s^2+2\zeta\omega_n s+\omega_n^2}$.

As it is necessary to consider the fraction $\dfrac{\mu_i\omega_{ni}s+\lambda_i\omega_{ni}^2}{s^2+2\zeta_i\omega_{ni}s+\omega_{ni}^2}$ that has a first-degree numerator where μ_i and λ_i are two constants, the corresponding impulse response becomes $h_i(t) = \dfrac{\omega_{ni}\exp(-\zeta_i\omega_{ni}t)}{\sqrt{1-\zeta_i^2}}\left[\lambda_i \sin\left(\omega_{ni}\sqrt{1-\zeta_i^2}\,t\right) + \mu_i\cos\left(\omega_{ni}\sqrt{1-\zeta_i^2}\,t + \arcsin\zeta_i\right)\right]$U($t$), of which the ZT is, according to the dictionary of the ZTs:

$$H_i(z) = \frac{\omega_{ni}}{\sqrt{1-\zeta_i^2}} \frac{\mu_i\sqrt{1-\zeta_i^2} + e^{-\zeta_i\omega_n T_e}\left[\lambda_i \sin(\omega_{ni}\sqrt{1-\zeta_i^2}T_e) - \mu_i\cos(\omega_{ni}\sqrt{1-\zeta_i^2}T_e - \arcsin\zeta_i)\right]z^{-1}}{1-2e^{-\zeta_i\omega_n T_e}z^{-1}\cos(\omega_{ni}\sqrt{1-\zeta_i^2}T_e) + e^{-2\zeta_i\omega_n T_e}z^{-2}}$$

The same procedure will be followed for a basic term of the first order whose samples are $y[k] = \exp(-\alpha k T_e)$, which would give $\mathrm{ZT}\{y(t)\} = \dfrac{1}{1-e^{-\alpha T_e} z^{-1}}$, while $\mathrm{LT}\{y(t)\} = \dfrac{1}{s+\alpha}$.

It is thus possible to carry out the transposition of the parameters on an element-wise basis in the partial fraction expansion of $H(s)$ to that of $H(z)$, and therefore to obtain $H(z)$ from $H(s)$ in the form of a rational fraction in z, assuming the conservation of the values of the sampled impulse response at times kT_e.

Example

Let the fourth-order type-II Chebyschev filter, inducing 40 dB of rejection in the stopband, with the angular frequency $\omega_c = 1$ rd/s, be determined using the function "cheb2ap" in MATLAB:

$$H(s) = K \frac{s^2+\omega_{z1}^2}{s^2+2\zeta_1\omega_{d1}s+\omega_{d1}^2} \frac{s^2+\omega_{z2}^2}{s^2+2\zeta_2\omega_{d2}s+\omega_{d2}^2},$$

with $\omega_{z1} = 1.0824$; $\omega_{z2} = 2.6131$; $\omega_{d1} = 0.5059$; $\omega_{d2} = 0.5591$; $\zeta_1 = 0.3383$; $\zeta_2 = 0.9025$ and K=1/100. The poles in the plane of the s variable are: $-0.1712 \pm 0.4761j$ and $-0.5045 \pm 0.2408j$. We choose the initial angular frequency $\omega_c = 1$ rd/s of the stopband equal to a quarter of the sampling angular frequency, that is $\omega_c / \omega_e = 0.25$ which gives $T_e = 1.5708$ s.

The bilinear transformation modified to obtain zero frequency distortion for ω_c is calculated with the coefficient $\omega_{bl} = \dfrac{\omega_c}{\tan\left(\pi \dfrac{\omega_c}{\omega_e}\right)} = \dfrac{1}{\tan(\pi/4)} = 1$ rd/s.

The variable s is replaced by $\omega_{bl}(z-1)/(z+1)$, which gives:

$$H(z) = K \frac{\omega_{bl}^2(z-1)^2+\omega_{z1}^2(z+1)^2}{\omega_{bl}^2(z-1)^2+2\zeta_1\omega_{d1}\omega_{bl}(z^2-1)+\omega_{d1}^2(z+1)^2} \frac{\omega_{bl}^2(z-1)^2+\omega_{z2}^2(z+1)^2}{\omega_{bl}^2(z-1)^2+2\zeta_2\omega_{d2}\omega_{bl}(z^2-1)+\omega_{d2}^2(z+1)^2}$$

or even:

$$H(z)=K\frac{(\omega_{bl}^2+\omega_{z1}^2)z^2+2z(\omega_{z1}^2-\omega_{bl}^2)+\omega_{bl}^2+\omega_{z1}^2}{(\omega_{bl}^2+2\zeta_1\omega_{d1}\omega_{bl}+\omega_{d1}^2)z^2+2z(\omega_{d1}^2-\omega_{bl}^2)+\omega_{bl}^2-2\zeta_1\omega_{d1}\omega_{bl}+\omega_{d1}^2}$$
$$\times\frac{(\omega_{bl}^2+\omega_{z2}^2)z^2+2z(\omega_{z2}^2-\omega_{bl}^2)+\omega_{bl}^2+\omega_{z2}^2}{(\omega_{bl}^2+2\zeta_2\omega_{d2}\omega_{bl}+\omega_{d2}^2)z^2+2z(\omega_{d2}^2-\omega_{bl}^2)+\omega_{bl}^2-2\zeta_2\omega_{d2}\omega_{bl}+\omega_{d2}^2}$$

that is in numerical values:

$$H(z)=K\frac{z^2(2.1716)+2z(0.1716)+(2.1716)}{z^2(1.5982)+2z(-0.7441)+(0.9136)}\frac{z^2(7.8284)+2z(5.8284)+(7.8284)}{2.3218z^2+2z(-0.6874)+0.3034}$$

and by normalizing the coefficients:

$$H(z)=4.58\times10^{-2}\frac{1+0.1580z^{-1}+z^{-2}}{1-0.9310z^{-1}+0.5716z^{-2}}\frac{1+1.4890z^{-1}+z^{-2}}{1-0.5922z^{-1}+0.1307z^{-2}}.$$

To achieve the transformation preserving the impulse response, we shall start by expanding $H(s)$ into partial fractions; in other words:

$$H(s)=\frac{1}{100}\frac{s^2+2(2-\sqrt{2})}{s^2+0.3423s+0.2559}\frac{s^2+2(2+\sqrt{2})}{s^2+1.0092s+0.3125}$$
$$=\frac{1}{100}-\frac{0.3977s+0.1415}{s^2+0.3423s+0.2559}+\frac{0.3842s+0.4821}{s^2+1.0092s+0.3125}.$$

The coefficients λ_1, μ_1, λ_2, μ_2, can be deduced by means of identification with the symbolic expression previously given for the last two fractions, then the corresponding z-transmittance:

$$H(z)=\frac{1+0.2248z^{-1}+0.9255z^{-2}}{1-1.1206z^{-1}+0.5841z^{-2}}\frac{1+3.676z^{-1}+0.3044z^{-2}}{1-0.8414z^{-1}+0.2049z^{-2}},$$

which comprises a pair of conjugate zeros and two real zeros.

The frequency responses are given in Figure 1.53 for the transmittance modulus.

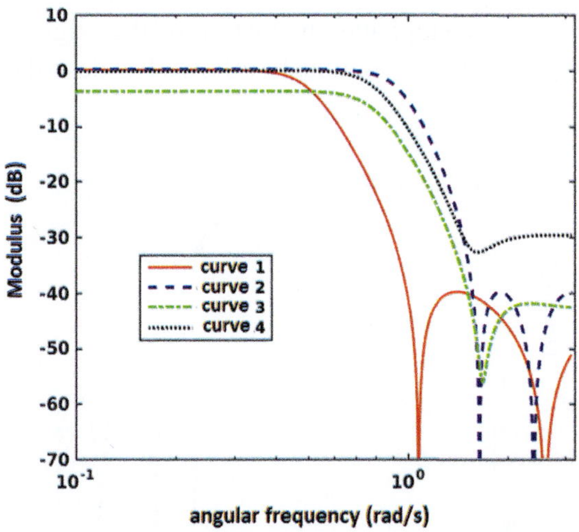

Figure 1.53. *Modulus of the transmittances; curve 1 = |H(jω)|; curve 2 = modulus of the ZT obtained by the bilinear approximation described in the text; curve 3 = modulus of the ZT obtained by the approximation which preserves the impulse response as described in the text; curve 4 = modulus of the ZT obtained by the approximation which preserves the impulse response, available in MATLAB ("impinvar") [MAT 94]. For a color version of this figure, see www.iste.co.uk/muret/electronics3.zip*

It should be noted that differences can be seen, mainly at the level of the cutoff frequency, that is shifted by a factor of approximately $\sqrt{2}$, and at the level of the transmittance zeroes, which are displaced or fewer in number. The shift in response no. 3 is probably due to rounding errors, but can be easily corrected by adding an amplification of about 3 dB, as illustrated in Figure 1.54. It can also be observed that the approximation to the preserved impulse response achieved in MATLAB [MAT 94] through the function "impinvar" (curve 4) gives a frequency response further away from the others than the one described above (curve 3), due to a less rigorous process for the computation of the numerator of $H(z)$. It especially suffers from a lack of attenuation in the stopband. In Figure 1.54, the response due to the bilinear approximation (curve 2) was replotted after modification of the pre-stressed angular frequency ω_{bl}, divided by 2 compared to the previous case, and the vertical shift of curve 3 rectified. Shifts that may appear are dependent on the type of the initial transmittance $H(s)$ and on the accuracy of the calculations. The slopes in the transition band nevertheless remain very comparable. This example also shows that it is possible to preserve both the impulse response and the stopband attenuation of the original transmittance, provided that the procedure be rigorously followed as described earlier.

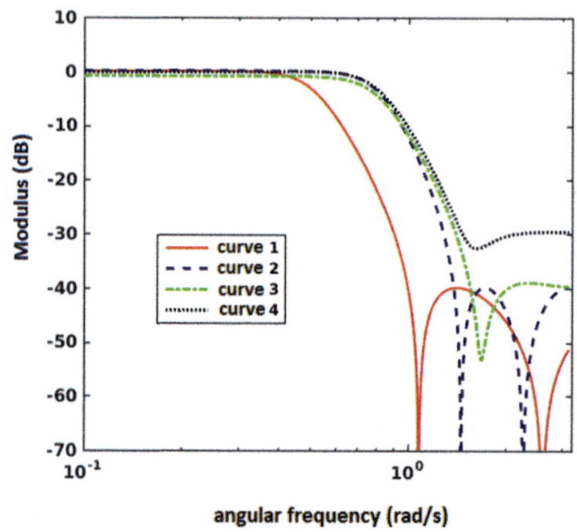

Figure 1.54. *Modulus of the transmittances; curve 1 = |H(jω)|; curve 2 = modulus of the ZT obtained by the bilinear approximation described in the text; curve 3 = modulus of the ZT obtained by the approximation which preserves the impulse response as described in the text; curve 4 = modulus of the ZT obtained by the approximation which preserves the impulse response, available in MATLAB ("impinvar") [MAT 94]. For a color version of this figure, see www.iste.co.uk/muret/ electronics3.zip*

1.4.6.3. *Types of sampled filters*

As previously discussed, sampled filters have a transmittance in the plane of the z variable that can be expressed in the form of a rational fraction:

$$H(z) = \frac{b_1 + b_2 z^{-1} + b_3 z^{-2} + \cdots + b_m z^{-m+1} + b_{m+1} z^{-m}}{a_1 + a_2 z^{-1} + a_3 z^{-2} + \cdots + a_n z^{-n+1} + b_{n+1} z^{-n}},$$

$$= \frac{(z - z_{z1})(z - z_{z2}) \cdots (z - z_{zm})}{(z - z_{p1})(z - z_{p2}) \cdots (z - z_{pn})} = \frac{N(z)}{D(z)}$$

where the z_{zi} (i =1 to m) and the z_{pi} (i =1 to n) are, respectively, the zeros and the poles of the transmittance in the plane of the z variable, whose map can be obtained in MATLAB using the function "zplane". The ZT of the input and output signals, respectively, $X(z)$ and $Y(z)$, of the filter are then connected by $Y(z) = H(z)\,X(z)$.

By developing this last expression using the first form of the transmittance multiplied in the numerator and in the denominator by z^i, and by taking the inverse transform, we get the recurrence equation between the samples of the input and of the output signal, following an inverted approach of that employed in sections 1.4.6.1 and 1.4.6.2:

$$\begin{aligned} & a_1 y[i] + a_2 y[i-1] + a_3 y[i-2] + \cdots + a_n y[i-n+1] + a_{n+1} y[i-n] \\ = \; & b_1 x[i] + b_2 x[i-1] + b_3 x[i-2] + \cdots + b_m x[i-m+1] + b_{m+1} x[i-m] \end{aligned}.$$

If all samples of the output signal after time iT_e are gathered in the right-hand side and assuming that the denominator is normalized by $a_1 = 1$, we can write the value of the current output sample as follows:

$$y[i] = a_2 y[i-1] + a_3 y[i-2] + \cdots + a_{n+1} y[i-n] + b_1 x[i] + b_2 x[i-1] + b_3 x[i-2] + \cdots + b_{m+1} x[i-m].$$

The second member includes both input and output samples in the more general case. Restricted to the case of a causal system, that is to say that assumes the samples of the input signal to be 0 for negative indexes, the sequence of output samples is written as follows:

$$\begin{aligned} & y[0] = b_1 x[0]; \\ & y[1] = a_2 y[0] + b_1 x[1] + b_2 x[0]; \\ & y[2] = a_2 y[1] + a_3 y[0] + b_1 x[2] + b_2 x[1] + b_3 x[0]; \\ & y[3] = \;\; \cdots. \end{aligned}$$

The calculation of output samples is exclusively achieved based on multiplication operations using the coefficients of the filter as well as addition, which is implemented in the case of digital filters as detailed further in the text. Here, the coefficients of the filter are none other than those of the variable z^{-1} in the numerator and the denominator of $H(z)$. Another computation is possible, because $Y(z) = H(z)\ X(z)$ implies the validity of the convolution product $y[i] = \sum_{k=0}^{+\infty} h[k]\, x[i-k]$ for a causal filter, this time using the samples of the impulse response which are the coefficients of the series expansion of $H(z)$ according to the decreasing powers of z starting from 0. For the FIR filters described hereafter, the two computation methods are identical.

The different types of filters can be distinguished according to the respective degrees m and n of the numerator and the denominator of $H(z)$. If $n = 0$, the denominator of $H(z)$ is a constant, and output samples are then calculated only by the addition and multiplication operations performed on m successive samples of the input signal. The impulse response is a sequence of a finite number of m samples and the filter is then called "finite impulse response (FIR) filter". This is also an average of successive samples of the input signal weighted by the coefficients b_k, which is known as moving because each output sample is calculated by way of the m previous input samples. For this reason, the filter is also known as "MA" (moving average). This type of filter is non-recursive because the output samples do not take part in the calculation.

If $m = 0$, the numerator of $H(z)$ is a constant, and the $n - 1$ previous output samples are used to calculate the current output sample. For this reason, the filter is known as recursive or autoregressive (AR). If $m \neq 0$ and $n \neq 0$, the filter is known as a MA and AR filter or ARMA. In both cases, the impulse response is infinite and these filters are called more generally as "IIR" filters.

IIR filters have the advantage of requiring an order comparable to that of analog filters to obtain an equivalent performance, and thereby exhibit maximal filtering efficiency for a minimal number of second-order cells. They have the disadvantage of having a frequency response which is characterized by a non-constant group velocity, which leads to more distortion when filtering pulses, compared to filters with constant group velocity. They can be subject to stability problems due to feedback (or recursive) loops inherent to their structure.

Conversely, FIR filters are instead unconditionally stable because of the absence of recursive loops (equivalent to a continuous-time feedback loop) but require an order equal to the number of first-order delay cells, much larger, to reach a filtering performance comparable to those of IIR filters. In most cases, they provide a linear frequency phase shift, or equivalently a constant group velocity. They have a number of transmittance zeros much more significant than IIR filters, which in general is an advantage in the stopband and as their name suggests, they show an impulse response of finite length and often devoid of sign change. This implies that the step response at any moment does not exceed the final value, which can prove to be a decisive advantage. Nevertheless, due to an order most of the time much higher than that of an IIR filter of comparable frequency response, the phase shift and thereby the delay of the output signal are also substantially increased compared to the input signal. These filters are very useful when it is desired to achieve a moving and weighted average over a series of samples. Although they are theoretically achievable using the analog technique of switched capacitors (see section 1.4.4), digital processing is better suited because it makes it possible to avoid voltage or current offsets and only rounding errors must be controlled. Last but not least, digital FIR filters can automatically increase the number of bits used for representing the output data compared to that of the input data since they include as many additive operations as the order of the filter.

1.4.6.4. *Synthesis of IIR filters*

There are several possible methods, which in most cases are based on calculation algorithms implemented by signal processing software or software specialized in filter synthesis. The first method starts from known analog prototypes for continuous-time filters, more specifically of the Butterworth, Chebyschev I or II or elliptical types, following the same criteria as those discussed in Chapter 2 of Volume 2 [MUR 17b], in order to adjust the frequency response of the filter to the desired template. The transposition in the domain of the z variable is then carried out using one of the two approximations previously described, either the bilinear transformation or the transformation preserving the impulse response, according to the priority constraint. The Bessel type is not utilized in practice in the domain of the z variable because much better performing linear-phase FIR filters do exist.

Other methods exist, in particular to build multi-band filters, namely presenting a transmittance modulus according to the frequency which is not similar to a conventional low-pass, band-pass, high-pass or band-stop filter template. These methods are called direct insofar as they do not resort to an approximation of the transformation of $H(s)$ into $H(z)$. On the other hand, they are all based on a minimization of the quadratic difference between a required response and the one calculated, in the spectral or time domain or still by using both, with at least a transformation from one domain to another. The Yule-Walker method (function “yulewalk” in MATLAB [MAT 94]) computes a transmittance based on a piecewise linear approximation of the curve of the modulus of the requested frequency response, but involving inverse transformations in order to achieve the minimization of the squared deviation between time responses. The “invfreqz” function does the same in MATLAB but from the complex frequency response (modulus and phase). The Prony method (function “prony” in MATLAB) delivers the coefficients of filter calculated based on the desired impulse response, by means of covariance computations; the filter is however not automatically stable. The Steiglitz-McBride method (function “stmcb” in MATLAB) achieves the same based on a time response of the output for an arbitrary input, provided that the numbers of samples are identical on input and on output. These last methods fall under the scope of statistics applied to signal processing, such as the construction of a predictor capable of delivering a current sample value based on the knowledge of previous samples, with a certain error margin. This is more particularly the case of the “lpc” function in MATLAB that calculates the coefficients of a linear prediction filter.

The architecture of these IIR filters implemented in switched-capacitor circuits can always be reduced to a cascading arrangement of second-order cells or sections. Independently of the filter, synthesis is greatly facilitated by the use of graphic subprograms present in software for applied mathematics and signal processing, in which we choose and capture the parameters and characteristics of the desired filter, such as “FDA” (“filter design and analysis tool”) in MATLAB, “xcos” in SciLab and so on. This kind of program is capable of directly delivering the coefficients of each of the second-order sections and of the possible first-order section. Another module is generally dedicated to graphically plotting the various types of response: frequency (modulus and phase), time (impulse, step), complex (zero and pole map), or group delay response. It allows the performance of the filter to be verified and can be complemented by a plot of the response to any input signal by performing its convolution with the impulse response.

1.4.6.5. *Synthesis of FIR filters*

Different synthesis principles can also be implemented for these filters. These are mainly: windowing; the equalization of the transmittance modulus in determined frequency bands; the minimization of the quadratic deviation over the whole of the frequency domain; the use of an algorithm capable of approximating an arbitrarily defined response, including a phase shift that may not be linear with frequency. The coefficients of FIR (or "MA") filters are defined in a unique way since the transmittance $H(z)$ is a series comprising a finite number of terms in powers of z^{-1}, which is also the numerator of $H(z)$, the denominator being equal to 1.

If we consider a sequence of M samples, windowing consists of transmitting without attenuation the median sample and in symmetrically attenuating the samples located on each of the sides using multiplicative factors decreasing according to a predetermined distribution $w[i]$ containing M non-zero coefficients too, which is also the sampled impulse response. If the impulse response of the ideal low-pass, band-pass, high-pass, band-stop or even multi-band filter, namely with unity transmittance in the bandwidth(s) and zero in the stopband, is $h_{id}[i]$, then the coefficients of the filter are simply $w[i]\, h_{id}[i]$ because due to the symmetry of $w[i]$ with respect to the median index, the convolution product is equivalent to a simple product. If $w[i]$ is constant and equal to 1, the effect in the case of the low-pass filter is simply that of a truncation, also corresponding to the impulse response of a comb filter (see Figure 1.21), whose transmittance is written as $H_{cb}(z) = \sum_{k=0}^{M} z^{-k} = \dfrac{1 - z^{-(M+1)}}{1 - z^{-1}}$. The frequency spectrum $H_{cb}(j\omega) = \dfrac{1 - \exp(-j(M+1)\omega T_e)}{1 - \exp(-j\omega T_e)} = \dfrac{\sin(\omega(M+1)T_e/2)}{\sin(\omega T_e/2)} \exp(-jM\omega T_e/2)$ is obtained by replacing z by $\exp(j\omega T_e)$. This filter is inefficient in terms of attenuation slope beyond the cutoff frequency at –3 dB, neighboring $\dfrac{\sqrt{6} f_e}{\pi M}$ based on a third-order expansion, because the maximal values of the lobes decrease as f^{-1}, like in a first-order filter.

We thus generally prefer employing windows giving a more pronounced cutoff slope and also retaining a phase factor linear with the frequency such as the one appearing at the end of the expression of $H_{cb}(j\omega)$.

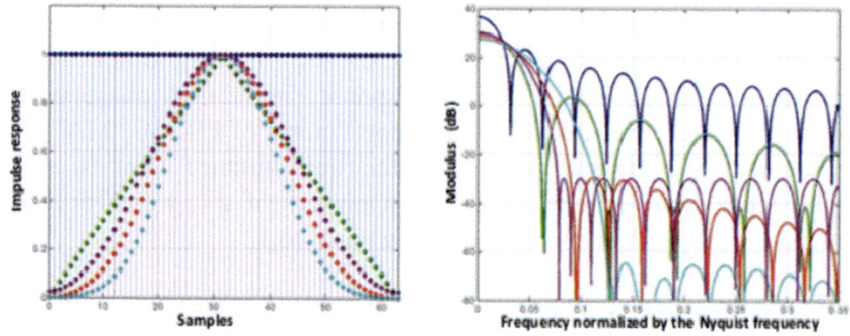

Figure 1.55. *Impulse (on the left) and frequency response (on the right) for FIR filters based on rectangular window or comb (in blue), Bartlett or triangular (in green), Chebyschev (in purple), Blackman (in red) and Blackman-Harris windows (in cyan color), with M = 64 coefficients. For a color version of this figure, see www.iste.co.uk/muret/electronics3.zip.*

A few examples are given in Figure 1.55 with the modules of $H(f)$ plotted according to the reduced frequency $f/(f_e/2)$ after normalization using the Nyquist frequency $f_e/2$ because, in the argument where the phase $\omega T_e = \dfrac{\pi f}{(f_e/2)}$ appears, this Nyquist frequency is naturally present in the denominator. Other windows like Gaussian, Hamming, Hann, Kaiser, Tukey and raised-cosine can be implemented. The choice is mainly achieved depending on the modification of the spectrum that they provide as they primarily operate to limit the discontinuity between the first and the last sample of a sequence, which is a necessary condition for using the discrete transform, which are then applied to compute a discrete spectrum. The reason has been exposed in section 1.1.3: discrete transforms assume that the signal is periodic with a period equal to the product of the number of samples and the sampling period T_e. The number of transmittance zeros is generally equal to that of the number of samples of the impulse response on the interval $[-f_e/2, f_e/2]$, with some exceptions (example of the triangular window, whose spectrum is the result of the squaring of the one of the rectangular window). Cutoff frequencies at –3 dB are of the same order as that of the comb filter, that is $\dfrac{2\sqrt{6}}{\pi M}\dfrac{f_e}{2} \approx \dfrac{1.56}{M}\dfrac{f_e}{2}$, resulting in 0.024

if referenced to the Nyquist frequency, in the case of the Figure 1.55 for which $M = 64$.

The second group of methods resorts to a minimization of the quadratic deviation between the calculated and desired transmittance moduli. This is the case when using functions "firls" or "remez" in MATLAB [MAT 94] by defining the frequency bands in which it is desirable to obtain the minimization of the gap. The "remez" algorithm is optimal in the sense that the ripple of the deviation due to the presence of transmittance zeros is equalized in the frequency bands under consideration (equiripple). On the other hand, a discontinuity still remains in general between the first and the last sample of the impulse response. The function "remezord" provides an estimate of the order of the Remez filter based on the normalized break of slope frequencies of the transmittance modulus and of the maximal deviation tolerated between desired and calculated moduli. As for IIR filters, the same graphics programs included in applied mathematics software and comprising a program dedicated to signal processing are capable of achieving the syntheses of FIR filters using the methods described above. The implementation using analog circuits is possible based on the circuit shown in section 1.4.4 and by further adding adders thereto that weigh each sample delayed by the coefficient of z^{-1} in the series $H(z)$, but at the cost of a very high number of delaying circuits, because the order of such filters is most of the time of several tens, which makes this solution scarcely used.

1.4.7. *Filtering and digital processing*

The operations that correspond to the mathematical operations previously analyzed in this chapter can be implemented on digital data. We can classify them into two categories, depending on the number of variables upon which the result depends: (1) in the first category, operations which result in a sampled signal only depending on time kT_e, or on the index of the time numbering k, and whose result can therefore be expressed by $y[k]$; (2) in the second category, operations whose final result depends on a second variable, other than the variable kT_e, while the latter acts as an internal variable in the computation. The first category may refer to computations continuously flowing, also referred to as being "in real time" or in a "streaming" way, especially when the processing system is part of a chain in which the samples follow each other at the rate of the sampling period. The application of such a computational method is slightly more problematic in the second case because it requires the storage of intermediate results. All can be performed by computerized systems, which can be either digital signal processors (DSPs), or functions developed based on those available in instrumentation and signal-processing software and made autonomous (run time). The benefits or requirements to operate by means of computer or micro-computing systems are justified by the possibility of working in floating point, eventually with complex numbers, and using memory storage. It then becomes possible to exploit all the possibilities of computation, of software development, of adapting to external constraints by changing computational parameters and of taking advantage of the memory size that these systems offer. This solution is well suited to the methods of the second category in which the storage in memory of input samples or samples after intermediate computations is essential. It is also well adapted if the constraint of the delay due to computational times is not an imperative priority, as is the case for the processing of signals already recorded in digital format. On the other hand, this solution may result in limitations on the execution speed and on the implementation of micro-programmed circuits of excessive complexity with regard to the task to be achieved, especially in the case where processing should be incorporated into a fixed integrated circuit and without adjustable parameters, as frequently met in embedded electronics. On the other hand, there is the possibility to perform the methods of the first category using standard logic circuits, without recourse to micro-programming or a system for storing data or intermediate results. This case is presented in the following section. The operations of the second category are then briefly discussed.

1.4.7.1. *Continuous-flow sample processing, known as "in real-time" or "streaming"*

The basic operation consists of multiplying the signed number D (the datum or signal) by an invariable constant C (the coefficient) of a given sign, and updating the result with a delay of one sampling period.

Let D be the mantissa of the decimal equivalent of a signed binary number and encoded into m bits: $D = d_{m-2}\times2^{m-2} + d_{m-3}\times2^{m-3} + \ldots + d_i\times2^i \ldots + d_1\times2 + db_0$; and d_{m-1} the sign bit. The coefficient C is a fixed, signed number and encoded into n bits, c_{n-1} being the sign bit and $c_{n-2}, c_{n-3}, \ldots c_2, c_1, c_0$ the weights of the powers of 2 of the mantissa in descending order, all known and either equal to 1 or equal to 0. In order to simplify the procedure, we will choose to give bits $c_{n-2}, c_{n-3}, \ldots c_j, \ldots c_2, c_1, c_0$ the weights of the absolute value of C.

The operational principle is shown in Figure 1.56. As described in Chapter 3 of Volume 1 [MUR 17a], the multiplication of D by C can be broken down into (1) the detection of the sign of the result by the function EXCLUSIVE OR, (2) the computation of the absolute value of D providing bits δ_i (section I, Figure 1.56); (3) the product of the absolute values of D and C implementing the shift of bits $\delta_{m-2}, \delta_{m-3}, \ldots \delta_i, \ldots \delta_1, \delta_0$ multiplied by c_j of j places to the left and then the addition of the resulting bits for each power of 2 (section II in 1.56); and (4) the encoding of the result as a signed number, illustrated in section III of Figure 1.56. All of these operations require exclusively combinational logic gates and switches. They are implemented in Figure 1.56 in the sets numbered I, II and III along with the representation of signed numbers with a half-scale shift. The function implemented in section IV by an arrangement of master–slave flip-flops allows the synchronization of the result with the clock after a delay of one sampling period. It is not necessary to retain all of the bits of the result, which are $m+n-1$ in total, and we can delete a number of less significant bits corresponding to a rounding error (i.e. from the right side in section III of Figure 1.56) according to the desired precision.

Exercise: carry out the modifications in the diagram that would allow the 2's-complement representation of signed numbers to be used.

For many FIR filters, the coefficients are all positive as assumed in Figure 1.56, and consequently it is possible to process unsigned signals. The multiplication operation by a coefficient is then reduced to section II only in Figure 1.56. This simplification of the complexity of the digital circuitry necessary for FIR filters compensates somewhat for the higher number of coefficients compared to IIR filters.

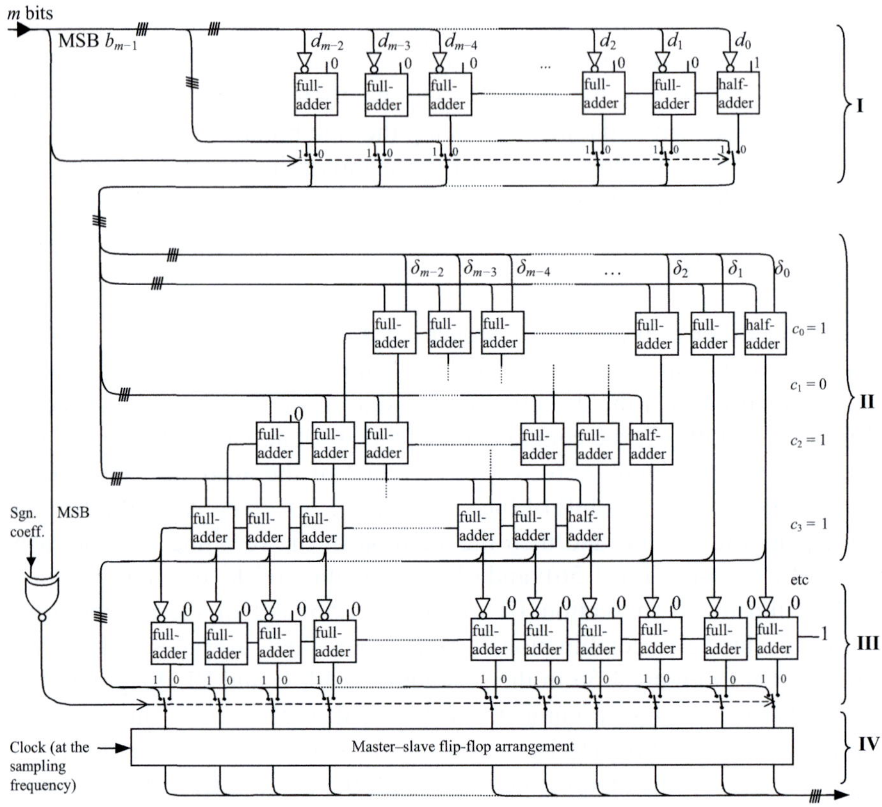

Figure 1.56. *Set of logical functions for the multiplication of a signed sample encoded into m bits by a signed constant (I, II and III) and delay of one sampling period of the result (IV)*

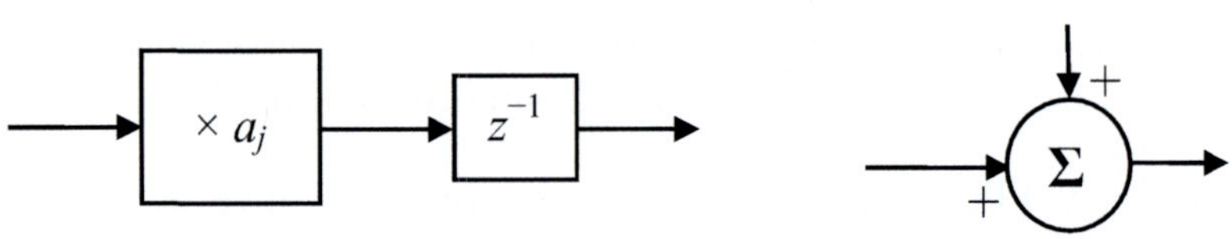

Figure 1.57. *Symbol of numerical multiplication operation by a coefficient and delay of one sampling period on the left and addition of numbers signed and encoded in binary on the right*

Another solution capable of performing the same operation consists of storing all the results of the product $C \times D$ in memory such as a lookup table; the bits of number D are then used as address bits. Such a solution is however likely to require more transistors than the previous one in the silicon wafer. Regardless of the computational method adopted, the basic operators can be symbolized as in Figure 1.57.

In order to achieve an IIR filter, different structures can be employed which make use of the basic operators previously detailed, making it possible to perform the multiplication by a constant, the delay of one sampling period, which is based on an assembly of master–slave flip-flops as in section IV of Figure 1.56, and the addition of signed numbers, summarized by symbols in Figure 1.57.

According to the given expression in 1.4.6.3 in the case where only a second-order transmittance is considered and by normalizing the first coefficient of the denominator of the transmittance to 1, the output is written as follows:

$$y[i] = a_2 y[i-1] + a_3 y[i-2] + b_1 x[i] + b_2 x[i-1] + b_3 x[i-2].$$

To implement a filter of order higher than 2, cascading several second-order sections proves to be a viable solution. If, on the contrary, it is desirable to implement the entire computation related to the full transmittance without breaking it down into second-order sections, the diagrams must be completed at the level of the dashed arrows (Figure 1.58 and 1.59) by adding the branches related to the coefficients b_4, b_5,... a_4, a_5 and so on to directly obtain the result given in section 1.4.6.3:

$$y[i] = a_2 y[i-1] + a_3 y[i-2] + \cdots + a_{n+1} y[i-n] + b_1 x[i] + b_2 x[i-1] + b_3 x[i-2] + \cdots + b_{m+1} x[i-m].$$

The system then operates according to an algorithmic sequence taking place following the basic operations arranged vertically in Figure 1.58 and 1.59, which shows other possible algorithms [MAT 94].

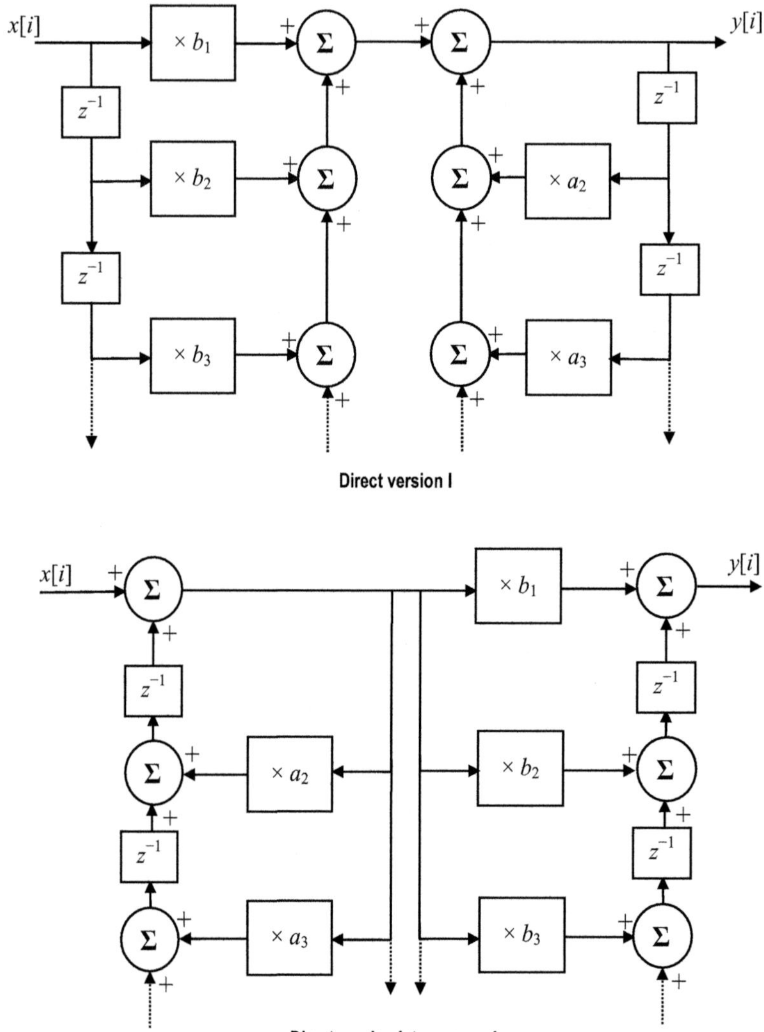

Figure 1.58. *Type I digital processing algorithms*

The same algorithms (Figure 1.58 and 1.59) are applicable for the implementation of FIR filters by removing the branches related to coefficients a_i and the corresponding adders. This is what is achieved to implement the improved decimation. In effect, the basic decimation operation, equivalent to subsampling, consists only of taking a sample among N according to a period NT_e if T_e was the initial period, but it incurs the loss of a huge amount of information since $N-1$ samples are ignored. In practice, this division of the sample frequency is associated with low-pass filtering allowing for both the recovery of the average value (or the sum) of samples during the interval NT_e and the attenuation of the entire portion of the spectrum corresponding to the fluctuations in the amplitude of the samples around this average value. At the same time, the filter implements the anti-aliasing function as a means, before downsampling, to remove in the ideal case or to strongly attenuate the portion of the spectrum that will fold after downsampling (see Figures 1.8 and 1.9). It may be noted that among the FIR filters described in section 1.4.6.5, the Blackman-Harris filter provides the strongest attenuation in stopband but at the expense of a frequency of the first transmission zero twice higher than that of the Chebyschev filter. Since cutoff frequencies are neighboring $\frac{1.56}{N}\frac{f_e}{2}$, the validation of the sampling theorem is not rigorously performed if f_e/N is adopted as output sampling frequency. It would be normally preferable to decrease them by a factor of 2 or 4, but this operation would require doubling or quadrupling N. And fourthly, the number of bits with which the result is encoded (often called "resolution") can be automatically increased following the property described for the basic operator of the multiplication by a coefficient at the beginning of this section (Figure 1.56): if the coefficient is encoded into n bits, which is often taken equal to $\log_2 N$, and the initial samples into m bits, the result can be encoded into $m+n-1$ bits at most. All of these properties are used in the sigma-delta converter described in Chapter 2.

The operations necessary for the convolution and correlation computations are similar and can be achieved with the basic operator of Figure 1.56, modified if the multiplicative factor is no longer a constant. In this case, it is enough to add AND gates on the path of the signals so as to either validate the value 0, or the value 1, for each bit c_{n-1}, c_{n-2}, c_{n-3}, ... c_2, c_1, c_0, depending on the value of the multiplication coefficient.

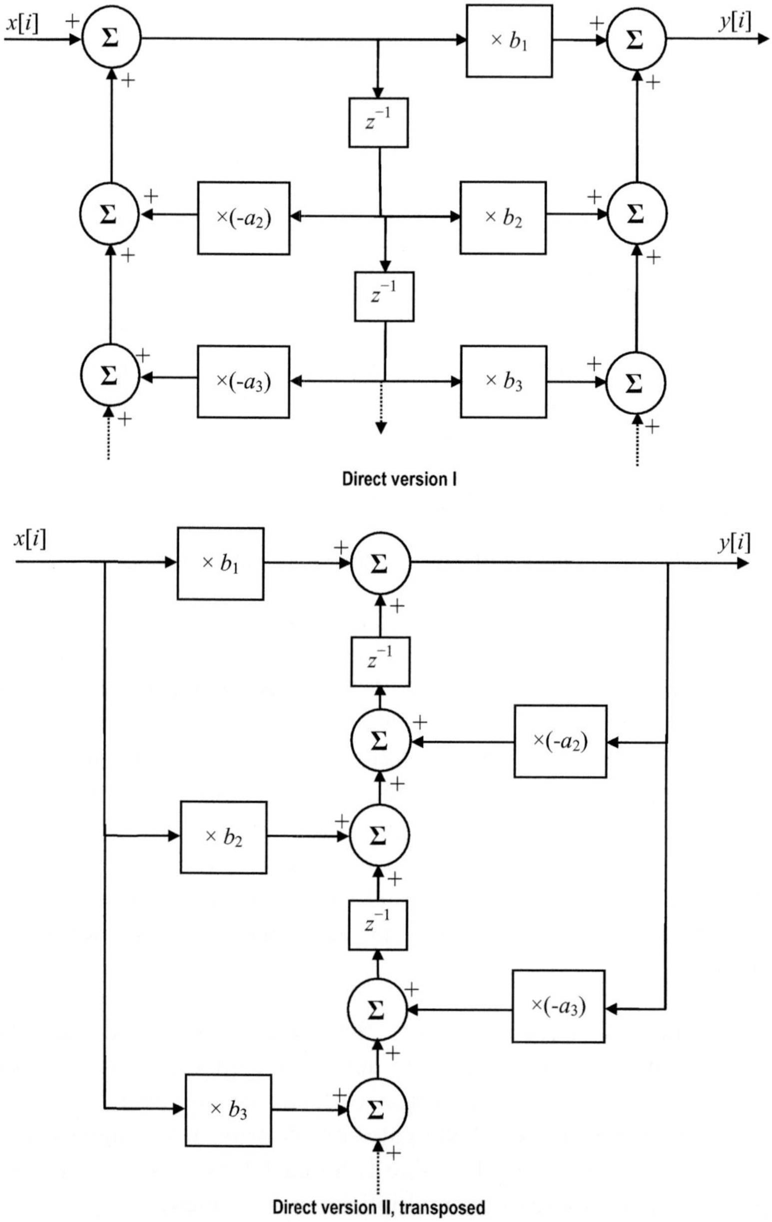

Figure 1.59. *Type II digital processing algorithms*

1.4.7.2. *Operations requiring the storage of intermediate results*

The operations that require the storage of intermediate results are interpolation or oversampling based on the application of the Shannon formula (see section 1.1.2) and transforms such as DFT, FFT, DCT, MDCT (see section 1.1.3) for the calculation of spectra or their inverses. They can be implemented in real time with the basic functions described in the previous paragraph, provided that functions are added for storing intermediate data that can be achieved by means of a number of shift registers equal to the number of bits utilized. The delay with which the result is output is obviously at least equal to the number of sampling periods needed to perform all operations, but this can still be considered as real-time processing based on operations carried out by an algorithmic system whose rate is imposed by the sampling clock. This option proves interesting if the computational system includes no adjustable parameters and must be fixed by way of cabled logic circuits (usually using a programmable circuit) in its final configuration to deliver a result in real time.

The other way to perform these computations is provided by microprocessors and associated peripherals, and more particularly by computational processors known as DSPs (digital signal processors) or by software running on a computer. Some software programs such as MATLAB and SciLab, are capable of both carrying out the requested operations in RAM in real time and of generating code that will be transferred to ROM in the DSP to turn the application autonomous, which enables development and testing prior to the final implementation. This is the path which is generally chosen if the application has to process pre-recorded data, without any significant constraint on the processing time, or if the computations must be achieved with adjustable parameters that require decision-based choices according to an algorithm involving loops and branches. This is, for example, the case for the search of the maximal correlation between a data set and predetermined templates.

1.5. Discrete-time state-space form

All systems studied in this chapter deal with sampled, analog or digital signals that can be represented by sequences, as described earlier in this chapter. The basic operations present in these systems can be equated to multiplications by a constant, delays of a sampling period T_e and summations or differences, following a measurement performed for each sampling period. A number of combinations of these basic operations complements the set of basic blocks necessary to define all recurrence relations only involving variables whose order differs by one unity only in the discrete-time domain. If we introduce as many variables (state variables) as necessary in order to obtain those first-order recurrence equations only, the whole electronic system can be described by a system of first-order sequences (or recurrence equations). It then becomes possible to adopt a state-space representation similar to that described in Chapter 1 of Volume 2 [MUR 17b], in the form of two matrix equations that allow us to define the relations between the input vectors $\mathbf{X}[k]$, $\mathbf{X}[k-1]$ and state variables $\mathbf{S}[k-1]$ at the instant numbered $k-1$, and $\mathbf{S}[k]$ and output vector $\mathbf{Y}[k]$ at the instant numbered k, with state matrix $\mathbf{A}$, control matrices $\mathbf{B}_0$ and $\mathbf{B}_1$, observation matrix $\mathbf{C}$ and feedthrough matrices $\mathbf{D}_0$ and $\mathbf{D}_1$:

$$(1) \quad \begin{cases} \mathbf{S}[k] = \mathbf{A}\mathbf{S}[k-1] + \mathbf{B} = \mathbf{A}\mathbf{S}[k-1] + \mathbf{B}_0\mathbf{X}[k] + \mathbf{B}_1\mathbf{X}[k-1] \\ \mathbf{Y}[k] = \mathbf{C}\mathbf{S}[k] + \mathbf{D} = \mathbf{C}\mathbf{S}[k] + \mathbf{D}_0\mathbf{X}[k] + \mathbf{D}_1\mathbf{X}[k-1] \end{cases}.$$

The derived state variable vector is here replaced by the state variable vector $\mathbf{S}[k]$ at the time numbered k, expressed as a function of vectors $\mathbf{S}[k-1]$, $\mathbf{X}[k]$ and $\mathbf{X}[k-1]$, which makes it possible to compute it in a recurrent way by repeating the matrix operation corresponding to the first equation as many times as necessary, starting from $k = 1$. Matrix computational software very easily performs this type of computation and is capable of obtaining the sequence of values taken by the samples of all the system variables, depending on the sequence of the sample values $\mathbf{X}[k]$ and $\mathbf{X}[k-1]$ imposed on the inputs.

If we calculate the eigenvalues μ_i and eigenvectors $\mathbf{U}_i$ of matrix $\mathbf{A}$, we can achieve the change of base that results in obtaining a diagonal matrix $\mathbf{M} = \mathbf{U}^{-1}\mathbf{A}\mathbf{U}$ which contains the eigenvalues μ_i as diagonal elements, and where the matrix $\mathbf{U}$, of elements u_{ij}, contains the eigenvectors $\mathbf{U}_j$ in the columns: $\mathbf{U}^{-1}\mathbf{S}[k] = \mathbf{M}\left(\mathbf{U}^{-1}\mathbf{S}[k-1]\right) + \mathbf{U}^{-1}\mathbf{B}$. If the elements of vectors $\mathbf{U}^{-1}$ $\mathbf{S}[k]$, $\mathbf{U}^{-1}$ $\mathbf{S}[k-1]$, $\mathbf{U}^{-1}\mathbf{B}$ are respectively defined by $\xi_i[k]$, $\xi_i[k-1]$ and η_i, this matrix equation is equivalent to the system of sequences $\xi_i[k] = \mu_i\, \xi_i[k-1] + \eta_i$ whose ZT gives $\xi_i(z) = \mu_i\, z^{-1}\, \xi_i(z) + \eta_i$ or even $\xi_i(z) = \dfrac{\eta_i}{1-\mu_i z^{-1}}$.

Therefore, we have a pole in the plane of the variable z every time that $z = \mu_i$.

Important conclusion: this last expression demonstrates that the eigenvalues of A are the poles in the plane of the z variable of the transmittance of the system because the eigenvalues are independent of the base of the matrix representation.

This last property also shows that there are an infinite number of state-space representations, obtained from a single one by all possible changes of base.

The definition of the two matrix equations of the system can be achieved either based on the recurrence equations of the system, such as those written for switched-capacitor circuits, or based on a block diagram, provided that it only involves first-order operators such as those that can be found in Figure 1.60. If there are higher-order z-transmittances or samples at times indices differing from more than one, it is important to introduce additional state variables so that in the end the system comprises first order sequences only. In Figure 1.60, the recurrence equations related to the first-order functional blocks already studied are indicated, taking into account that the operator z^{-1} induces a delay of one sampling period corresponding to the decrease of one unity of the order number of the sample. It may be necessary to solve partial systems of recurrent sequences to isolate a state variable.

In practice, it is often easier to use a second form of the state-space form in which $\mathbf{S}[k-1]$ and $\mathbf{S}[k]$ are interchanged. As a matter of fact, the determination of the expressions of $\mathbf{S}[k-1]$ based on $\mathbf{S}[k]$ is much more consistent with the reality of systems that sequentially process samples and systematically introduce a delay. In other words, $\mathbf{S}[k]$ is the cause and $\mathbf{S}[k-1]$ the consequence, which is thus evaluated more easily from $\mathbf{S}[k]$ and not the other way around.

It can then be written as:

$$(2)\quad \begin{cases} \mathbf{S}[k-1] = \mathbf{A}'\mathbf{S}[k] + \mathbf{B}' = \mathbf{A}'\mathbf{S}[k] + \mathbf{B}'_0\mathbf{X}[k] + \mathbf{B}'_1\mathbf{X}[k-1] \\ \mathbf{Y}[k] = \mathbf{C}\mathbf{S}[k] + \mathbf{D} = \mathbf{C}\mathbf{S}[k] + \mathbf{D}_0\mathbf{X}[k] + \mathbf{D}_1\mathbf{X}[k-1] \end{cases}$$

By multiplying by $\mathbf{A}$ the first matrix equation, we get:

$$\mathbf{A}\,\mathbf{S}[k-1] = \mathbf{A}\,\mathbf{A}'\mathbf{S}[k] + \mathbf{A}\mathbf{B}' = \mathbf{A}\,\mathbf{A}'\mathbf{S}[k] + \mathbf{A}\mathbf{B}'_0\mathbf{X}[k] + \mathbf{A}\mathbf{B}'_1\mathbf{X}[k-1].$$

It can be seen that it is necessary to impose **A A'** = **I** (where **I** is the unit matrix), that is **A'** = $\mathbf{A}^{-1}$, the inverse matrix of **A**, to return to the first form:

$$\mathbf{S}[k]=\mathbf{A}\,\mathbf{S}[k-1]-\mathbf{A}\,\mathbf{B}'=\mathbf{A}\,\mathbf{S}[k-1]-\mathbf{A}\,\mathbf{B}'_0\mathbf{X}[k]-\mathbf{A}\,\mathbf{B}'_1\mathbf{X}[k-1],$$

after inversion of **A'** to obtain **A**. We also have $\mathbf{B}_0=-\mathbf{A}\mathbf{B}'_0$ and $\mathbf{B}_1=-\mathbf{A}\mathbf{B}'_1$.

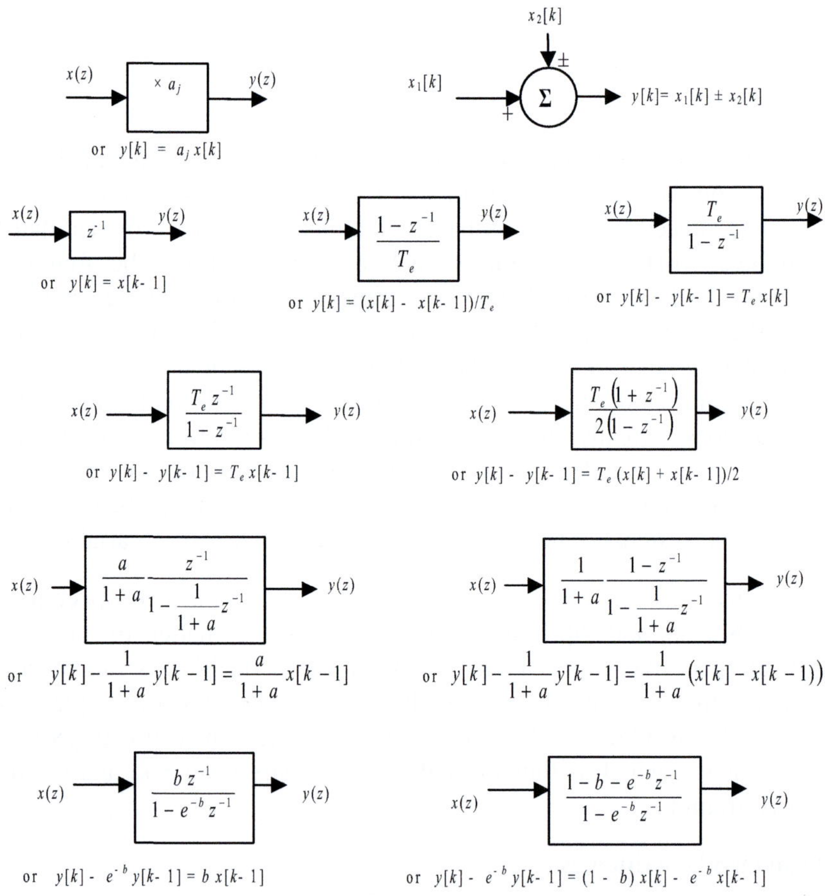

Figure 1.60. *Recurrence equations related to the functional blocks of the first-order in z above, from top to bottom and from left to right: multiplication by a coefficient, adder-subtractor, delay of one sampling period T_e, derivation, integration using the right rectangle method, integration using the left rectangle method, integration using the trapezoid method, first order switched-capacitor passive low-pass and high-pass filters, first-order low-pass and high-pass analog filters (for notations a and b, see respectively section 1.4.3.1 and Figure 1.52)*

Provided that the eigenvalues are independent of the chosen base, the comparison of those of **A** and **A'** in a specific base for each representation shows that the eigenvalues of **A'** are the inverse of the poles of the system in the plane of the z variable. In the event the notation of the second matrix equation does not appear to be immediate, it is possible to resort to the stationarity property that is always implicitly satisfied in these systems and allows for all the indices of the sequence to be modified by adding or subtracting a unity therefrom. Furthermore, it is useful to calculate the inverse of matrix **A** or **A'** in order to ensure that it is not singular. In fact, a singular matrix indicates that the number of state variables that have been chosen is too high and that these variables are not all independent, which must lead to the revision of the system of recurrence equations.

In conclusion, the representation based on state variables allows us to model all the stationary sampled systems, analog or digital, verifying the sampling theorem, and to obtain the evolution of the sequences of samples of the state variables and on output, usually by employing software for systems of order >2. The calculation method is simpler than in the continuous-time state-space representation as it only requires that stages be concatenated consisting, after each evaluation of **S**[k] and **Y**[k], in inserting the result in the second member of the matrix system (1) from k=1 and to iterate. The input variables are the initial data that make it possible for each step to completely evaluate the second member. It is even possible to address the case in which the sampling period is not constant (frequency modulated signals) by adapting the coefficients of the matrices at every computation step and this process can also be applied to systems including blocks having a nonlinear response. In both cases, some matrix coefficients become dependent of state variables and require that they be reexamined and eventually modified at each computational time step. The simulation and analysis of the responses of these systems can be performed by means of software programs (MATLAB, SciLab, etc.) based on a block diagram built using predetermined functional blocks whose parameters have to be detailed (Simulink, xcos, etc.), and that are able to simulate any elementary function that may exist. Finally, by solving one of the two systems of equations, which requires that **A** or **A'** be diagonalized, it is possible to express an output variable after replacement of the state variables according to the input variables at different times. Then by taking the ZT of this recurrence relation, we can deduce the transfer function of this system in the plane of the z variable and the eigenvalues of **A** or **A'** give direct access to the poles of the transmittance.

1.6. Exercises

1.6.1. *Switched-capacitor first-order high-pass filter*

Type P and I switches (MOS transistor-based) are respectively closed (switched on) when the command is 1 and open (switched off) when it is zero, corresponding to P in phase with integer times $(k-1)T_e$, kT_e and so on, and I in inverted phase; in other words closed in synchronism with half-integer times $(k-1/2)T_e$, $(k+1/2)T_e$ and so forth. They are "break before make" switches, in such a way that opening happens just before closing.

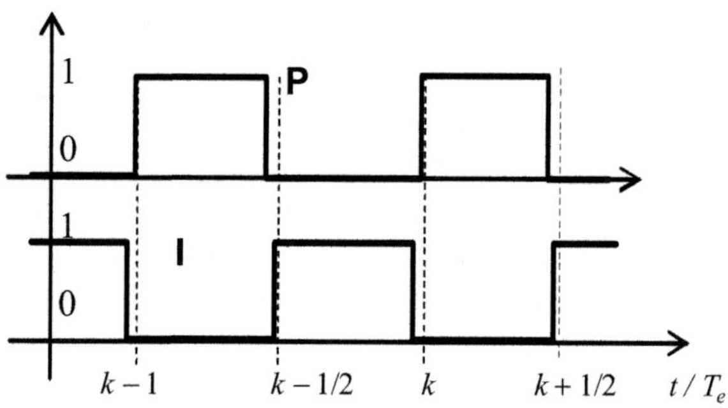

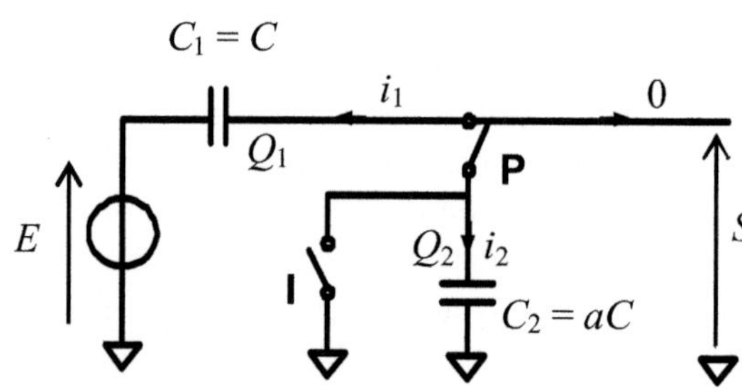

1) Write the equations connecting the charges Q_1 and Q_2 to voltages E and S for the circuit at time indices $k-1$, $k-1/2$ and k: $Q_1[k-1]$, $Q_2[k-1]$, $Q_1[k-1/2]$ and so on, as a function of $E[k-1]$, $S[k-1]$, $E[k-1/2]$ and so forth. Also write the relations of charge conservation during the transitions at time indices $k-1/2$ and k. Deduce, by eliminating the quantities evaluated at $k-1/2$, the recurrence relation between $E[k-1]$, $S[k-1]$, $E[k]$ and $S[k]$.

2) Determine the transmittance in z of the assembly $T_2(z) = S(z)/E(z)$ from the ZT of the previous equation. The parameter $b = \dfrac{1}{1+a} < 1$ can be used.

3) Determine the samples of the step response $S_U(z)$, that is when $E(z) = \dfrac{1}{1-z^{-1}}$, by carrying out the division of the numerator by the denominator of $S_U(z) = T_2(z)E(z)$. What is the initial value and toward what final value does it converge?

4) For sinusoidal signals $z = \exp(j\omega T_e)$ and if $\omega T_e << 1$, it can be written at the first order that $z^{-1} = 1 - j\omega T_e$. In this case, determine $T_2(j\omega)$, the nature of this transmittance and the associated time constant. What must be the condition upon a in order for this approximation to be still valid at the cutoff angular frequency? If it is satisfied, is the sampling theorem respected at the cut-off frequency?

Answers:

1) The equations are of the Q=CV and charge conservation types:

a) At time index $k - 1$:

$Q_1[k-1] = C(S[k-1] - E[k-1])$;

$Q_2[k-1] = aC\, S[k-1]$.

b) At time k–1/2, switch P switches off (opening) and therefore, the charge Q_1 does not change:

$Q_1[k - 1/2] = Q_1[k - 1]$.

In addition:

$Q_1[k-1/2] = C(S[k-1/2] - E[k-1/2])$

and $Q_2[k - 1/2] = 0$, because C_2 is shortly circuit by the switch I which is closing (switching on).

Therefore it can be inferred that

$$S[k-1/2] - E[k-1/2] = S[k-1] - E[k-1].$$

c) At time index k:

$$Q_1[k] = C\,(S[k] - E[k]);$$

$$Q_2[k] = aC\, S[k].$$

At the transition $k-1/2 \rightarrow k$:

$$\int_{k-1/2}^{k} (i_1 + i_2)\, dt = 0 = Q_1[k] - Q_1[k-1/2] + Q_2[k] - Q_2[k-1/2]$$

involving $0 = S[k] - E[k] + a\,S[k] - (S[k-1] - E[k-1])$ from which the recurrence relation is deduced:

$$(1 + a)\, S[k] - S[k-1] = E[k] - E[k-1].$$

2) The ZT gives: $(1 + a)\, S(z) - z^{-1}\, S(z) = E(z) - z^{-1}\, E(z)$

$$(1 + a - z^{-1})\, S(z) = (1 - z^{-1})\, E(z).$$

By establishing $b = \dfrac{1}{1+a}$: $T_2(z) = \dfrac{S(z)}{E(z)} = b\,\dfrac{1-z^{-1}}{1-bz^{-1}}$.

3) With $E(z) = \dfrac{1}{1-z^{-1}}$, we get: $S_u(z) = \dfrac{1}{1-z^{-1}}\, b\, \dfrac{1-z^{-1}}{1-bz^{-1}} = b\,\dfrac{1}{1-bz^{-1}}$.

According to the initial value theorem, the sample $S_U[0]$ is obtained from $S_U(z)$ when $z \rightarrow \infty$, which gives $S_U[0] = b$.

By division:

$$S_U(z) = b\left(1 + b\,z^{-1} + b\,z^{-2} + b\,z^{-3} + \cdots\right) = b + b^2\, z^{-1} + b^3\, z^{-2} + b^4\, z^{-3} + \cdots$$

corresponding to samples $S_U[0] = b$; $S_U[1] = b^2$; $S_U[2] = b^3$; … ; $S_U[k] = b^{k+1}$.

Given that $b < 1$, this series tends to zero and thus reproduces well the step response of a high-pass filter:

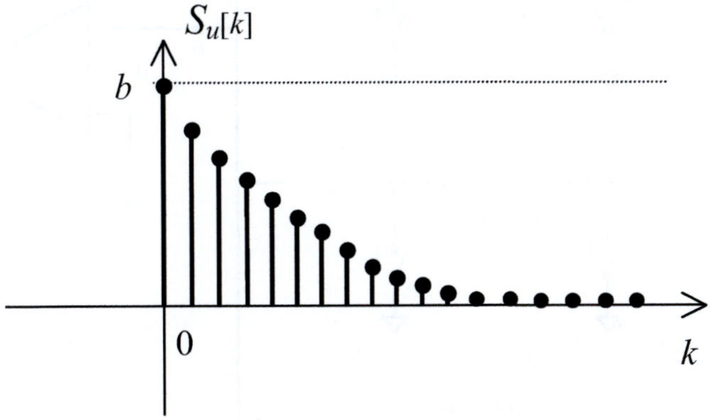

4) For sinusoidal signals $z = \exp(j\omega T_e)$ and if $\omega T_e << 1$, it can be written as $z^{-1} = 1 - j\omega T_e$ at first order with a relative error smaller than 2% if $\omega T_e < 4$; hence:

$$T_2(j\omega) = b\frac{j\omega T_e}{1 - b + j\omega T_e b} = \frac{j\omega \frac{T_e}{a}}{1 + j\omega \frac{T_e}{a}}.$$

that is, the transfer function of a high-pass filter of time constant T_e/a. However, in order for this approximation to be applicable to $\omega_c = a/T_e$, it is necessary that $\omega_c T_e << 1$ thereby $a << 1$. In this case $f_c = \frac{\omega_c}{2\pi} << \frac{1}{\pi}\frac{f_e}{2}$ and the sampling theorem is respected.

1.6.2. *Basic switched-capacitor-based filter operator (IIR) using an ideal operational amplifier.*

Same functioning conditions of the P- and I-type switches (MOS transistor-based) as in exercise 1.6.1.

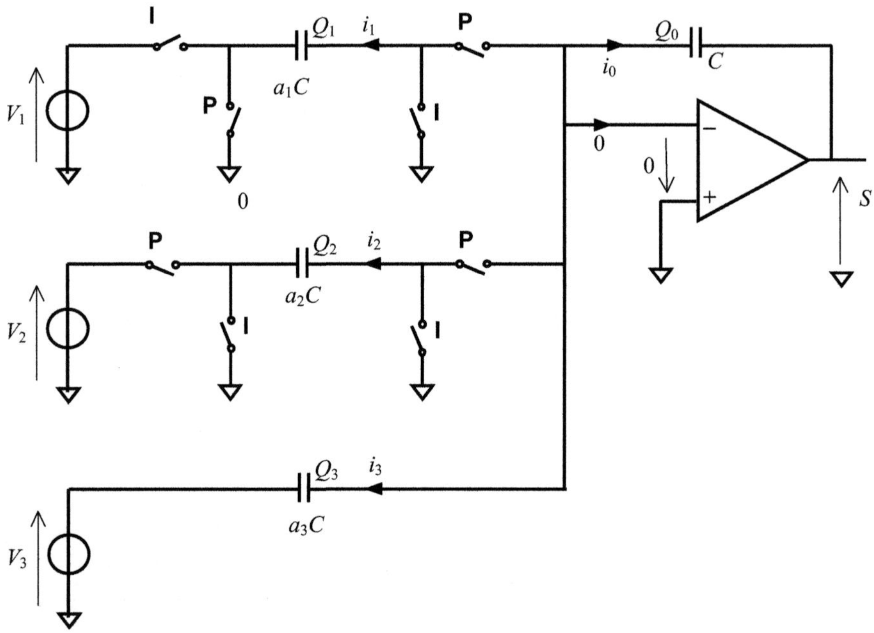

1) Same question as in exercise 1.6.1 for the assembly above.

2) Determine $S(z)$ according to $V_1(z)$, $V_2(z)$, $V_3(z)$, z^{-1}, a_1, a_2, a_3 by taking the ZT of the previous recurrence equation. What should the zero coefficients be to obtain an integrator without time delay? A time-delayed integrator?

3) Which operation is transmittance $\dfrac{z^{-1}}{1-z^{-1}}$ equivalent to on the circle $z = \exp(j\omega T_e)$ in the approximation $\omega T_e \ll 1$. Determine the samples of the step response for this transmittance.

4) Deduce from previous question 2 the transmittance in the plane of the z variable for assemblies nos. 1 and 2 of the following pages. Build a diagram of the charge transfer transmittances in each branch. Give the conditions to ensure that (1) the coefficients of the second order transmittance be all positive; (2) only the z^{-1} coefficients of the denominator be negative.

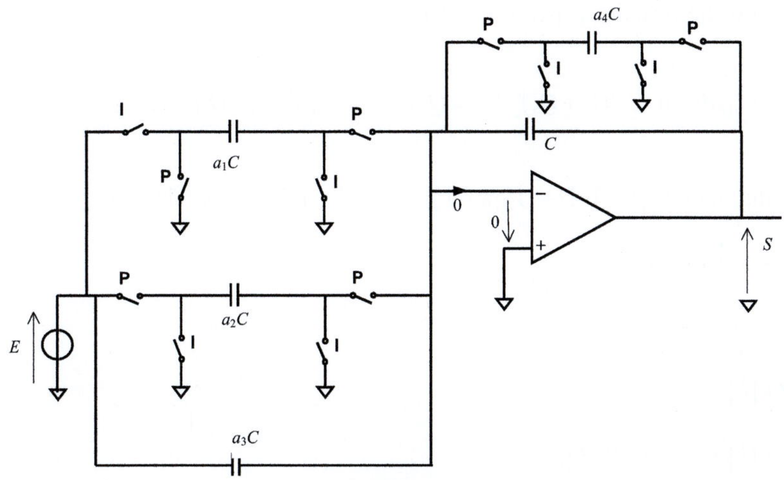

Assembly no. 1

Assembly no. 2: see Figure 1.49.

Answers:

1) The equations are of the Q=CV and charge conservation types:

a) At time index $k-1$:

$Q_0[k-1] = -C\,S[k-1]$,

$Q_1[k-1] = 0$,

$Q_2[k-1] = -a_2\,C\,V_2[k-1]$,

$Q_3[k-1] = -a_3\,C\,V_3[k-1]$.

b) At time index $k-1/2$:

$Q_0[k-1/2] = -C\,S[k-1/2]$,

$Q_1[k-1/2] = -a_1\,C\,V_1[k-1/2]$,

$Q_2[k-1/2] = 0$,

$Q_3[k-1/2] = -a_3\,C\,V_3[k-1/2]$.

During the transition from $k-1$ to $k-1/2$

$$\int_{k-1}^{k-1/2}(i_3+i_0)\,dt=0=-(Q_3+Q_0)_{k-1}^{k-1/2}=Q_3[k-1]+Q_0[k-1]-Q_3[k-1/2]-Q_0[k-1/2]$$

involving: $a_3\,(\,V_3[k-1/2]-V_3[k-1]\,)\;+\;S[k-1/2]-S[k-1]=0$.

c) At time k:

$Q_0[k]=-C\,S[k]$,

$Q_1[k]=0$,

$Q_1[k]=-a_2\,C\,V_2[k]$,

$Q_3[k]=-a_3\,C\,V_3[k]$.

During the transition from $k-1/2$ to k

$$-\int_{k-1/2}^{k}(i_0+i_1+i_2+i_3)\,dt=0=-(Q_0+Q_1+Q_2+Q_3)_{k-1/2}^{k}$$

involving:

$$0=S[k]-S[k-1/2]-a_1\,V_1[k-1/2]+a_2(V_2[k]-0)+a_3\,(V_3[k]-V_3[k-1/2]).$$

Nonetheless, according to the transition from $k-1$ to $k-1/2$,

$a_3\,V_3[k-1/2]+S[k-1/2]=a_3\,V_3[k-1]+S[k-1]$, giving after substitution:

$$S[k]=S[k-1]+a_1\,V_1[k-1/2]-a_2\,V_2[k]-a_3(\,V_3[k]-V_3[k-1]\,).$$

2) For the sampling theorem to be applied, the voltages must not vary much between the half sampling periods. It may thus be considered that $V_1[k-1/2]=V_1[k-1]$, which amounts to considering integer index times only for the observation of voltages (if this approximation is not to be carried out, the factor $z^{-1/2}$ will be used instead of z^{-1} for this term in the ZT). The ZT of the recurrence equation then gives:

$$S(z)=z^{-1}S(z)+a_1\,z^{-1}\,V_1(z)-a_2\,V_2(z)-a_3\,(1-z^{-1})\,V_3(z)\,.$$

Hence finally:

$$S(z) = a_1 \frac{z^{-1}}{1-z^{-1}} V_1(z) - a_2 \frac{1}{1-z^{-1}} V_2(z) - a_3 V_3(z) \cdot$$

For a non-delayed integrator, branches 1 and 3 are removed ($a_1 = 0$, $a_3 = 0$) and for a delayed integrator, the branches 2 and 3 are removed ($a_2 = 0$, $a_3 = 0$).

3) If $z = \exp(j\omega T_e)$ and $\omega T_e << 1$, we get $z \approx 1 + j\omega T_e$ and $z^{-1} \approx 1 - j\omega T_e$ at first order; then $\frac{z^{-1}}{1-z^{-1}} = \frac{1}{z-1} \approx \frac{1}{j\omega T_e}$ corresponds to integration in the time domain (when $\omega T_e << 1$, there is also not much difference between $\frac{z^{-1}}{1-z^{-1}}$ and $\frac{1}{1-z^{-1}}$). More accurately, $\frac{1}{1-z^{-1}} = \frac{z}{z-1} \approx \frac{1+j\omega T_e}{j\omega T_e}$ is equivalent to the inverse transmittance of a high-pass filter, but this only makes sense for $\omega << 1/T_e$ so as to ensure that $\omega T_e << 1$ and in this approximation there is no difference between both integrators.

For the step response, we multiply by the ZT of the unit step $\frac{1}{1-z^{-1}}$ and the division of the numerator by the denominator according to the increasing powers of z^{-1} is performed. Concerning the transmittance $\frac{1}{1-z^{-1}}$, the ZT of the step response is thus $\frac{1}{\left(1-z^{-1}\right)^2} = \frac{1}{1-2z^{-1}+z^{-2}}$, which gives $1 + 2z^{-1} + 3z^{-2} + 4z^{-3} + \ldots + (n+1)z^{-n}$ corresponding to a series of successive samples 1 ; 2 ; 3 ; … ; n+1, from time zero. For the response $\frac{z^{-1}}{\left(1-z^{-1}\right)^2}$, the delay theorem simply has to be applied and the whole previous series is delayed by one sample step, or more precisely 0 ; 1 ; 2 ; …n from time zero. Alternatively, a vertical translation of the samples can also be achieved to move from one series to the other by observing that $\frac{z^{-1}}{1-z^{-1}} = \frac{1}{1-z^{-1}} - 1$.

4) From the base assembly, it is possible to add or remove branches of the V_1 or V_2 or V_3 type, to re-loop one or two inputs onto the output, and to insert a second stage following it by re-looping onto the first stage. This is what is achieved in the next proposed assemblies.

For the first assembly, $V_1 = V_2 = V_3 = E$ and $V_4 = S$ are applied to a branch of the same kind as that upon which V_2 is applied; therefore:

$$S(z) = a_1 \frac{z^{-1}}{1-z^{-1}} E(z) - a_2 \frac{1}{1-z^{-1}} E(z) - a_4 \frac{1}{1-z^{-1}} S(z) - a_3 E(z) \cdot$$

Namely: $S(z)\left[1 + \dfrac{a_4}{1-z^{-1}}\right] = \dfrac{a_1 z^{-1} - a_2 - a_3\left(1 - z^{-1}\right)}{1-z^{-1}} E(z) \cdot$

Hence finally: $S(z) = \dfrac{-\left(a_2 + a_3\right) + \left(a_1 + a_3\right) z^{-1}}{1 + a_4 - z^{-1}} E(z)$

will give at will a low pass with $a_2 = a_3 = 0$; or a high-pass with $a_1 = a_2$ or still a transfer function that combines both of them.

We transform Assembly no. 2 into a diagram of charge transfer functions dependent on z sketched below, from which are derived two equations of conservation of the zero currents on the minus inputs of the operational amplifiers:

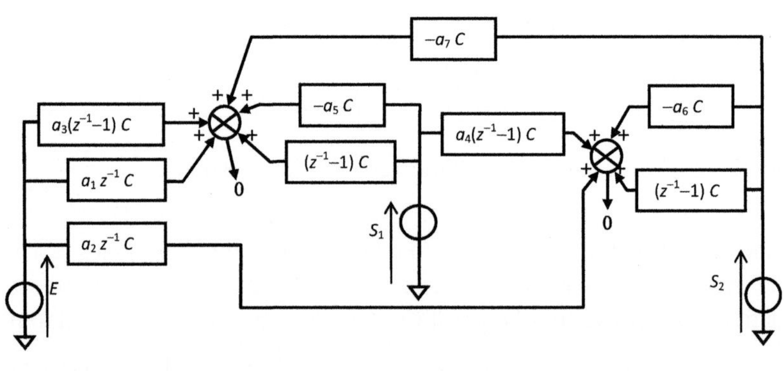

$$\left[a_1 z^{-1} + a_3\left(z^{-1} - 1\right)\right]E(z) + \left[-a_5 + \left(z^{-1} - 1\right)\right]S_1(z) - a_7 S_2(z) = 0$$

$$a_4\left(z^{-1} - 1\right)S_1(z) + a_2 z^{-1}E(z) + \left[-a_6 + \left(z^{-1} - 1\right)\right]S_2(z) = 0$$

By eliminating $S_1(z)$, we get:

$$\frac{S_2(z)}{E(z)} = \frac{a_3 a_4 + \left[a_2(1 + a_5) - a_1 a_4 - 2a_3 a_4\right]z^{-1} + \left[a_4(a_1 + a_3) - a_2\right]z^{-2}}{(1 + a_5)(1 + a_6) - a_4 a_7 + (a_4 a_7 - a_5 - a_6 - 2)z^{-1} + z^{-2}}$$

and by eliminating $S_2(z)$, we get:

$$\frac{S_1(z)}{E(z)} = -\frac{a_3(1 + a_6) + \left[a_2 a_7 - (a_1 + 2a_3 + a_1 a_6 + a_3 a_6)\right]z^{-1} + (a_1 + a_3)z^{-2}}{(1 + a_5)(1 + a_6) - a_4 a_7 + (a_4 a_7 - a_5 - a_6 - 2)z^{-1} + z^{-2}}.$$

a) To obtain positive coefficients in the first transmittance, it is necessary that $a_2(1 + a_5) > a_1 a_4 + 2a_3 a_4$ and $a_4(a_1 + a_3) > a_2$ in the numerator, which also implies summing the two inequalities member-wise, $a_2 a_5 > a_3 a_4$ and for the denominator $(1 + a_5)(1 + a_6) > a_4 a_7$ and $a_4 a_7 > a_5 + a_6 + 2$, which also implies $a_5 a_6 > 1$. For the second transmittance, it is necessary that $a_2 a_7 > a_1 + 2a_3 + a_1 a_6 + a_3 a_6$. All these conditions are achievable as well as reciprocal conditions. For example:

b) with $(1 + a_5)(1 + a_6) > a_4 a_7$, the first term of the denominator remains positive and the coefficient of z^{-1} becomes negative if $a_5 + a_6 + 2 > a_4 a_7$. It is also necessary that $a_5 a_6 > 1$ and it suffices that $a_4 a_7$ be decreased sufficiently to satisfy these conditions.

1.6.3. *Delay operator with offset correction and FIR filtering*

Given the following circuit in which V_0 represents the voltage offset of the amplifier:

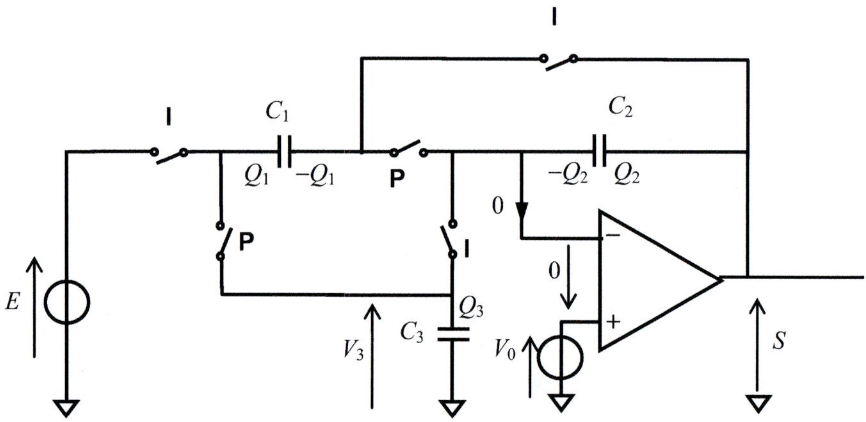

Since the capacitance C_2 is constantly in the feedback loop onto the operational amplifier assumed to be ideal (zero input current), the latter imposes a zero voltage between its – and + inputs, and therefore the offset voltage V_0 on the minus input. It should be noted that, since its output is an ideal voltage source, the conductor formed by the right armatures of capacitors C_1 and C_2 when I is closed (switched on) does not constitute an isolated assembly. The switches P and I are closed (switched on) when the command is 1 and they are opened (switched off) when it is 0, following the sequence already indicated in exercise 1.6.1. The transmittance is to be determined by writing the relations between charges and voltages at times $(k-1)T_e$, $(k-1/2)T_e$ and kT_e, which will be denoted by $Q_1[k]$, $Q_2[k]$, $S[k]$, $Q_1[k-1/2]$, $Q_2[k-1/2]$, $S[k-1/2]$, $E[k-1/2]$ and so on, in the manner detailed below:

1 a) At time $(k-1/2)T_e$, write the relation between the charge $Q_1[k-1/2]$, C_1 and voltages $S[k-1/2]$ and $E[k-1/2]$.

b) What happens at time $(k-1/2)T_e$ for C_2? Deduce thereof the relation between $Q_2[k-1/2]$ and $Q_2[k-1]$ then the one between $S[k-1/2]$ and $S[k-1]$.

c) Since the sampling theorem can be applied and the voltage E is updated only once per period T_e, we will perform the approximation $E[k-1/2] = E[k-1]$. Deduce $Q_1[k-1/2]$ from the previous relations with respect to C_1 and voltages $S[k-1]$ and $E[k-1]$. What is the voltage at the terminals of C_3?

2 a) At time kT_e, what value does charge $Q_1[k]$ assume?

b) Write the charge conservation on the armatures of capacitors C_1-C_3 and C_1-C_2 which are in contact during the transition at instant kT_e as well as the relations between charges and potentials in C_2. Deduce thereof $V_3[k] - V_0$ and $Q_1[k]$ as a function of C_1, C_2, C_3, $S[k]$ and $S[k-1]$.

c) According to the previous equations, establish the relation between $S[k]$, $S[k-1]$ and $E[k-1]$ and show that it is correctly independent of V_0.

3) We define $a = C_2/C_1$. Establish the expression of the transmittance $T(z) = S(z)/E(z)$ for any a.

4) By taking $C_3 = C_1$, determine C_2 so that $T(z)$ is equivalent to a simple delay of the input signal of one period T_e. What is the function of M identical circuits arranged in cascade?

5) It is assumed that the sum of the M output voltages is performed with an identical weighting for each one (comb filter). Establish the ZT of the output of the adder $S_{cb}(z)$ according to $E(z)$ and express the transmittance in a compact form with respect to z, and then with respect to ωT_e in the case of a sinusoidal input of frequency $f << f_e/2 = 1/(2T_e)$.

Answers:

1) a) At time $(k-1/2)T_e$, $Q_1[k-1/2] = C_1\,(E[k-1/2] - S[k-1/2])$,

b) and the capacitor C_2 is isolated because of the zero input current of the operational amplifier; therefore, $Q_2[k-1/2] = Q_2[k-1]$; $S[k-1/2] = S[k-1]$ and $Q_2[k-1/2] = C_2(S[k-1] - V_0)$.

c) The potential differences at the terminals of C_1 and C_3 are such that

$Q_1[k-1/2] = C_1\,(E[k-1] - S[k-1])$ and $V_3[k-1/2] = V_0$; hence $Q_3[k-1/2] = C_3 V_0$.

2) a) At time kT_e, closing the two switches P located between C_1 and C_3 yields

$$Q_1[k] = C_1(V_3[k] - V_0).$$

b) The left armature of C_1 and the upper armature of C_3 form an isolated single system, with the right armature of C_1 at voltage V_0. Therefore,

$$Q_1[k] + Q_3[k] - Q_1[k - 1/2] - Q_3[k - 1/2] = 0;$$

hence $C_1(V_3[k] - V_0) + C_3(V_3[k] - V_0) - C_1(E[k - 1] - S[k - 1]) = 0.$

Therefore $V_3[k] - V_0 = \dfrac{C_2}{C_1 + C_3}(S[k] - S[k-1])$ and $Q_1[k] = \dfrac{C_1 C_2}{C_1 + C_3}(S[k] - S[k-1])$.

In addition, the armatures of C_1 and C_2 are put in contact through the switch P and are isolated; therefore $Q_1[k] - Q_1[k - 1/2] + Q_2[k] - Q_2[k - 1/2] = 0$. Since the operational amplifier imposes voltage V_0 on the left armature of C_2, we also get $Q_2[k] = C_2(S[k] - V_0)$.

c) We deduce by substituting in the second charge conservation equation: $\dfrac{C_1C_2}{C_1+C_3}(S[k]-S[k-1]) - C_1(E[k-1]-S[k-1]) + C_2(S[k]-V_0) - C_2(S[k-1]-V_0) = 0$,

from which it can be inferred that

$$\left(\frac{C_1C_2}{C_1+C_3}+C_2\right)S[k]-\left(\frac{C_1C_2}{C_1+C_3}+C_2-C_1\right)S[k-1]=C_1E[k-1].$$

3) The ZT is therefore:

$$\left(\frac{C_2}{C_1+C_3}+a\right)S(z)-\left(\frac{C_2}{C_1+C_3}+a-1\right)z^{-1}S(z)=z^{-1}E(z)$$

or:

$$S(z)=\frac{z^{-1}E(z)}{\left(\dfrac{C_2}{C_1+C_3}+a\right)-\left(\dfrac{C_2}{C_1+C_3}+a-1\right)z^{-1}}.$$

4) In order for $S(z) = z^{-1} E(z)$, it is necessary that $\left(\dfrac{C_2}{C_1+C_3}+a\right)=1=C_2\left(\dfrac{2C_1+C_3}{(C_1+C_3)C_1}\right)$. Hence the condition $(C_1+C_3)C_1 = C_2(2C_1+C_3)$. If we take $C_3 = C_1$, $C_2 = \dfrac{2C_1}{3}$ and $a = \dfrac{C_2}{C_1} = \dfrac{2}{3}$ have to be imposed.

The input signal is thus delayed by one sample period, and by arranging M stages in cascade, we can obtain the input signal delayed by kT_e on output of the k-th stage, with $k \in [0\,, M]$, like in a shift register for a binary logic signal.

5) We have $S_{cb}(z) = E(z)\sum_0^M z^{-k} = E(z)\dfrac{1-z^{-(M+1)}}{1-z^{-1}}$ and by substituting z^{-1} by $e^{-j\omega Te}$: $S_p(j\omega) = E(z)\dfrac{1-\exp(-j(M+1)\omega T_e)}{1-\exp(-j\omega T_e)}$ namely a transmittance in sinusoidal regime $\dfrac{\sin((M+1)\pi f T_e)}{\sin(\pi f T_e)}\exp(-jM\pi f T_e)$ provided that $f < f_e/2$ be satisfied, of the low-pass type with a "gain" $M+1$ when $f \to 0$, which is the result of the summation of the input and of the M outputs.

The transmittance is 0 for $f_k = \dfrac{k}{(M+1)T_e} = \dfrac{kf_e}{M+1}$ (the integer $k \in [0\,, M/2]$) and the complex exponential factor involves a delay $\dfrac{MT_e}{2}$.

1.6.4. *Phase-locked loops*

The algebraic phase difference φ between the signals $s_1(t)$ and $s_2(t)$ can be considered as the difference of the instantaneous phase of either one, that is $\varphi_1(t) - \varphi_2(t)$, determined from the time interval Δt between rising edges. The measurement is carried out by the charge or discharge of a capacitor C_0 with constant current during the fraction Δt of the period by the circuit shown on the next page, itself controlled by the sequential logic circuit described in the course (Figure 1.25).

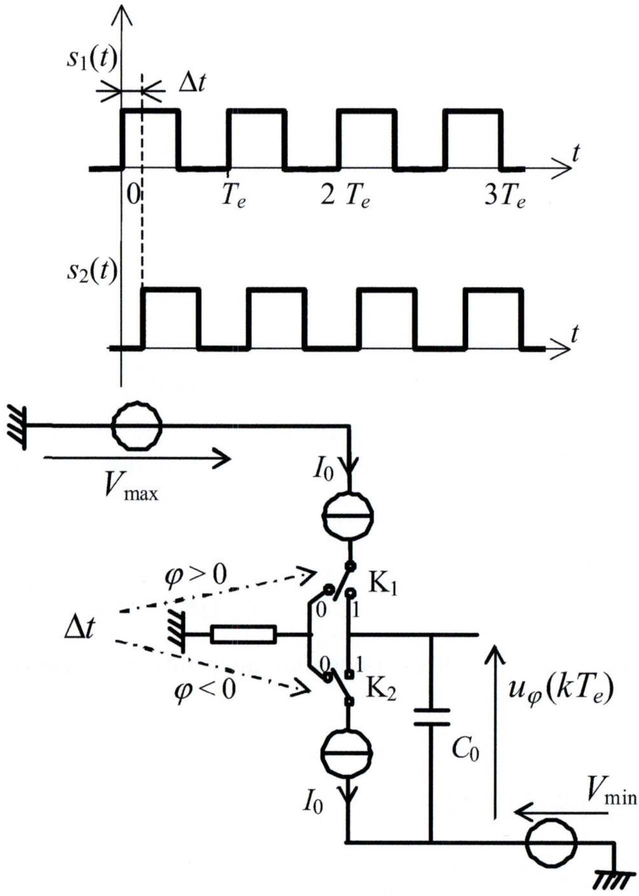

We make the approximation that the positive or negative increase of the voltage $u_\varphi(t)$ at the terminals of C are due to a current $i_\varphi(t)=\dfrac{I_0}{2\pi}\left[\varphi_1(t)-\varphi_2(t)\right]$. Both of these changes can then be considered as continuous-time signals regardless of the frequency of $s_1(t)$ and $s_2(t)$.

1) In its simplest version, the block diagram of the PLL is obtained by completing the previous device with a corrector comprising an impedance $Z_0(s)$ crossed by the current $I_\varphi(s)$, as the LT of $i_\varphi(t)$, with Z_0 incorporating at least the capacitance C_0 referenced to the voltage V_{max} instead of V_{min} in the previous diagram, a VCO with transmittance K_f (K_f real), a frequency divider by N (N integer), the functional block $A_3(s)$ making it possible to transform

frequency into phase shift and a unit return loop. The block diagram is sketched in the following figure.

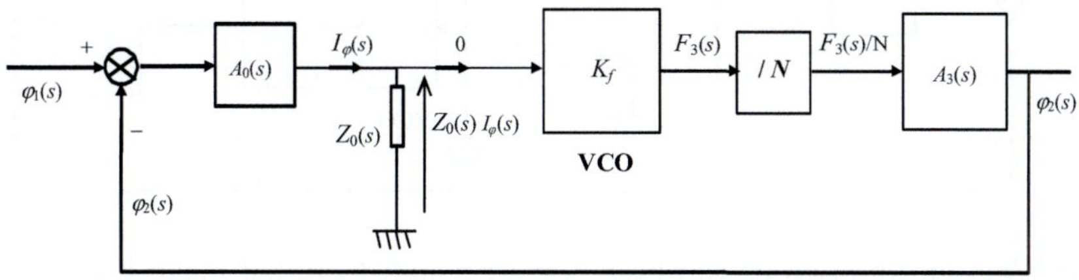

Reminder of the time relation between instantaneous phase and frequency:

$$\varphi(t) = 2\pi \int_0^t f(t')\, dt' .$$

a) Determine the loop transmittance (direct chain only) then that of the closed-loop system $T_1(p)$ and the nature of Z_0 for the closed-loop transmittance to be of the second order with a natural angular frequency $\omega_{n1} = \sqrt{f_2 \Delta f_2}$, maintaining the notations of the course and a damping coefficient ζ_1 to be determined as a function of the system parameters.

b) Discuss the possibilities of applying the conditions indicated in the course concerning the parameters determining the dynamic response, especially a damping coefficient $\frac{1}{\sqrt{2}}$ if Δf_2 is equal to $f_{2\max}/10$ and if the maximal modulation frequency is $f_{\mathrm{Mmax}} = f_{2\max}/100$ and $f_{2\max}$= 100 MHz. Is the cutoff frequency of the low-pass filter based on Z_0 acceptable?

2) We modify the diagram of the loop filter by substituting Z_0 with the only capacitance C_0 connected to the bias supply $V_{\max}$ so that the LT of the voltage variations at the terminals of C_0 be equal to $-I_\varphi(s)/(C_0\, s)$ and by adding a corrective filter built by an inverting amplifier whose gain is $-\dfrac{Z_2(s)}{Z_1(s)}$ (the necessary follower between the capacitance C_0 and the inverting amplifier as shown in Figure 1.29 in the course is included in $A'_1(s)$) as indicated in the following block diagram:

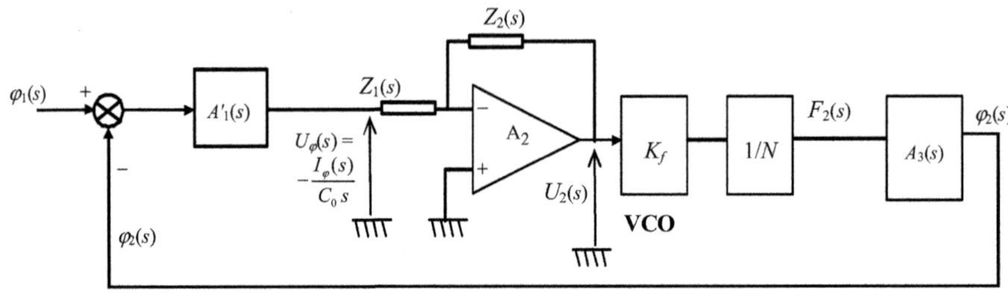

a) What are the two possibilities for $Z_1(s)$ and $Z_2(s)$ that make it possible to obtain a closed-loop transmittance of the second-order low-pass type? We will choose the one comprising no inductance.

b) Determine the closed-loop transmittance $H_2(s)$ and its parameters ω_{n2} and ζ_2.

c) Same question as in 1) b) if we want the natural frequency 10 times greater than f_{Mmax}. Evaluate the gain-bandwidth product necessary for the operational amplifier.

3) To accelerate the response to a unit step (step response), a derived correction with time constant τ_d is added to the phase lead correction already present in the loop filter.

a) What would be the shape of the derivative of the actual output of the phase-shift detector (solid line in Figure 1.24 of the course)? It will be however assumed that the VCO is sensitive to the average value of the control voltage for a period of the signal and reacts to the derivative of the continuous-time signal (dashed line in the figure of the course), which is equivalent to assuming that the transmittance of this corrector is simply $-\tau_d s$. What is $Z_1(s)$ composed of?

b) Determine the closed-loop transmittance $H_3(s)$ and its parameters ω_{n3} and ζ_3. We will use the following notations: $\alpha = \tau_d/(R_2 C_1)$; $\beta = R_2/R_1$ and $x = \omega_{n1} R_1 C_1$. Show that the minimal value of ζ_3 is obtained for $x = \sqrt{\frac{1}{\alpha\beta}}$. By adopting this value, demonstrate that it is necessary to take $\alpha = 1$ to obtain $\zeta_3 = \sqrt{\frac{1}{2}}$. Deduce β thereof to get $\omega_{n3} = 8\omega_{n1}$ then the various time constants. Finally, calculate the maximal asymptotic gain A_m of the bandpass term of $H_3(j\omega)$.

Assuming that with $\zeta_3 = 1/\sqrt{2}$, the step response is damped after a delay of $2/f_{n3}$, where $f_{n3} = \omega_{n3}/(2\pi)$, how many periods of an input signal at 95 MHz are necessary to recover the locking?

Answer:

1) a) According to the block diagram, the loop transmittance $T_1(s)$ is equal to $\dfrac{I_0\, Z_0(s)\, K_f}{N\, s}$; hence the closed-loop transmittance:

$H_1(s) = \dfrac{T_1(s)}{1+T_1(s)} = \dfrac{I_0\, Z_0(s)\, K_f}{N\, s + I_0\, Z_0(s)\, K_f}$, which becomes of the second order if $Z_0(s)$ has a denominator that is proportional to s. However, if only the capacitance C_0 was inserted, which is necessary to implement the integral phase-shift detector function, the damping coefficient would be 0 (oscillator!). To avoid instability, we will include a resistance R_0 in series: $Z_0(s) = \dfrac{1}{C_0 s} + R_0$. Hence:

$$H_1(s) = \frac{1 + R_0 C_0 s}{1 + R_0 C_0 s + \dfrac{N C_0}{I_0 K_f} s^2}.$$

This is second-order transmittance with a natural angular frequency $\omega_{n1} = \sqrt{\dfrac{K_f I_0}{N C_0}}$ and a damping coefficient $\zeta_1 = \dfrac{1}{2} R_0 C_0 \sqrt{\dfrac{K_f I_0}{N C_0}}$. Note: even in the case where $N \neq 1$, the system can be reduced to a system without any frequency divider and in which the new VCO would have a transmittance K_f/N.

b) By adopting the relation

$\dfrac{K_f}{N}\dfrac{I_0}{C_0} = \dfrac{K_f}{N}\dfrac{I_0 T_e}{C_0 T_e} = \dfrac{\Delta f_2 (V_{max} - V_{min})}{(V_{max} - V_{min}) T_e} = f_{2\,max}\, \Delta f_2$, we get $\omega_{n1} = \sqrt{f_{2\,max}\, \Delta f_2}$ and $\zeta_1 = \dfrac{1}{2} R_0 C_0 \sqrt{f_{2\,max}\, \Delta f_2}$.

Numerical application: $f_{n1} = \frac{\omega_{n1}}{2\pi} = 5$ MHz and it is thus necessary that $R_0 C_0 = \frac{\sqrt{2}}{3.15 \times 10^7} = 45$ ns. The first-order filtering of the current I_0 is thus achieved with this time constant, or more specifically a 3.5 MHz cutoff frequency that is not adjustable. However, since the harmonics of the output current of the phase-shift detector are at least at frequencies $f_{2\min} = 90$ MHz and multiples, the filtering is absolutely correct. The frequency f_{n1} is five times greater than $f_{M\max}$, which also seems reasonable. On the other hand, if we wanted to accelerate the loop response even further, it would be necessary to increase ω_{n1} by increasing I_0 (or decreasing C_0), which would have the disadvantage of limiting the range of linear operation of the phase-shift detector at a value less than 2π.

2) a) We now have:

$$T_2(s) = \frac{I_0}{2\pi C_0 s} \frac{Z_2(s)}{Z_1(s)} \frac{K_f}{N} \frac{2\pi}{s} = \frac{I_0 Z_2(s) K_f}{C_0 Z_1(s) N s^2} = \frac{Z_2(s)\,\omega_{n1}^2}{Z_1(s) s^2}.$$

To obtain a non-zero closed-loop damping coefficient, we have the choice between Z_2 including an inductance and a resistance in series and $Z_1 = R_1$, or alternatively $Z_2 = R_2$ and a capacitance C_1 in series with a resistance R_1 for Z_1. We thus choose the second option, devoid of inductance, which gives: $H_2(s) = \frac{T_2(s)}{1+T_2(s)} = \frac{\omega_{n1}^2 R_2}{Z_1(s)\, s^2 + \omega_{n1}^2 R_2} = \frac{1}{1 + \frac{s}{R_2 C_1 \omega_{n1}^2} + \frac{R_1}{R_2 \omega_{n1}^2} s^2}$, namely a natural angular frequency $\omega_{n2} = \omega_{n1}\sqrt{\frac{R_2}{R_1}} = \sqrt{\frac{R_2}{R_1}}\sqrt{f_{2\max}\,\Delta f_2}$ and a damping coefficient $\zeta_2 = \frac{1}{2}\sqrt{\frac{R_2}{R_1}} \frac{1}{R_2 C_1 \omega_{n1}} = \frac{1}{2C_1\sqrt{R_2 R_1 f_{2\max} \Delta f_2}}$, adjustable independently of the sensitivity of the phase-shift detector. If we want $\omega_{n2} = 2\,\omega_{n1}$ ($\Rightarrow f_{n2} = 10\, f_{M\max} = 10$ MHz) and $\zeta_2 = \frac{1}{\sqrt{2}}$, then $R_2 = 4R_1$ and $R_1 C_1 = 11{,}2$ ns, corresponding to 14 MHz for the cutoff frequency of the filtering of the current $I_\varphi(s)$. The closed-loop transmittance is now purely of the low-pass type.

3) a) The derivative of the actual output of the phase-shift detector is a sequence of impulses smaller than the period except if the loop is completely unlocked. In order for us to continue within the context of a linear approximation, it will be assumed that the VCO reacts to the average value over a period of this signal, which is equivalent to considering the transmittance $-\tau_d\, p$. It is theoretically obtained by adding a capacitance $C_d = R_2 / \tau_d$ in parallel to the R_1-C_1 series network, hence the new expression of $Z_1(s)$ in the computation of $U_2(s)$ that follows (for the practical solution, see section 1.3.3).

b) Now we have $U_2(s) = -\dfrac{I_\phi(s)}{C_0}\left[\dfrac{R_2}{R_1 + \dfrac{1}{C_1 s}} + \tau_d s\right]$ and $T_3(s) = \dfrac{\omega_{n1}^2}{s^2}\left[\dfrac{R_2}{R_1 + \dfrac{1}{C_1 s}} + \tau_d s\right]$ for the loop transmittance, from which it is deduced in closed loop:

$$H_3(s) = \frac{T_3(s)}{1+T_3(s)} = \frac{1 + \dfrac{R_1 C_1 \tau_d}{R_2 C_1 + \tau_d} s}{1 + \dfrac{R_1 C_1 \tau_d + 1/\omega_{n1}^2}{R_2 C_1 + \tau_d} s + \dfrac{R_1 C_1 s^2}{(R_2 C_1 + \tau_d)\omega_{n1}^2}} = \frac{1 + A_m \dfrac{s}{\omega_{n3}}}{1 + 2\zeta_3 \dfrac{s}{\omega_{n3}} + \dfrac{s^2}{\omega_{n3}^2}}$$

with $\omega_{n3} = \omega_{n1}\sqrt{\dfrac{R_2 C_1 + \tau_d}{R_1 C_1}} = \omega_{n1}\sqrt{\beta}\sqrt{1+\alpha}$; $A_m = \sqrt{\dfrac{1}{\beta}}\dfrac{\alpha\beta R_1 C_1 \omega_{n1}}{\sqrt{1+\alpha}}$

and

$$\zeta_3 = \frac{1}{2}\sqrt{\frac{R_1 C_1}{R_2 C_1 + \tau_d}}\left[\omega_{n1}\tau_d + \frac{1}{\omega_{n1} R_1 C_1}\right] = \frac{1}{2}\sqrt{\frac{R_1}{R_2}}\sqrt{\frac{1}{1+\alpha}}\left[\alpha\omega_{n1} R_2 C_1 + \frac{1}{\omega_{n1} R_1 C_1}\right]$$

or $$\zeta_3 = \frac{1}{2}\sqrt{\frac{1}{(1+\alpha)\beta}}\left[\alpha\beta x + \frac{1}{x}\right]$$

The derivative of the bracket with respect to x is 0 for $x = \sqrt{\dfrac{1}{\alpha\beta}}$; the bracket is then equal to $2\sqrt{\alpha\beta}$, that is $\alpha = 1$ to obtain $\zeta_3 = \zeta_{3\min} = \sqrt{\dfrac{1}{2}}$.

To obtain $\omega_{n3} = 8\,\omega_{n1}$, it is then necessary that $\beta = 32$, that is $R_2 = 32\,R_1$.

The time constants are inferred from $\omega_{n1} = 3.14\times10^7$ rad/s, giving $R_1\,C_1 = 5.6$ ns and $\tau_d = R_2\,C_d = R_2\,C_1 = 180$ ns. We then get a natural frequency $f_{n3} = 8\,f_{n1} = 40$ MHz, which significantly approaches $\sqrt{f_{2\max}\,f_{2\min}} = 94.8$ MHz while maintaining an optimal damping factor. The maximal asymptotic gain of the bandpass part of $H_3(s)$ is equal to $A_m = \sqrt{\dfrac{\alpha}{1+\alpha}} = \dfrac{1}{\sqrt{2}}$, which provides a contribution close to that of the low-pass term, thus improving the speed of the step response and therefore the flexibility of the loop. The step response is damped in 2/(40 MHz), or 50 ns, which represents only about five periods of the input signal to restore the lock.

1.6.5. *Sampled models of the PLL*

By means of initially removing the correction proportional to the phase shift, it has been shown in the previous exercise (question 2) that an effective loop filter was that of the Figure 1.29 with $Z_2 = R_2$ and Z_1 comprising a resistance R_1 in series with a capacitance C_1.

1) a) Under these conditions, determine the transmittance $G(s)$ of this loop filter, the shape of its asymptotic Bode diagram and the type of correction that it achieves.

b) Determine what switched-capacitor-based assembly can be achieved by adopting one of the inputs of the basic first-order filtering operator (switched-capacitor-based) of exercise 1.6.2 as input to the filter and by connecting the other two inputs to its output, such that to obtain a filter whose frequency response $G_1(z)$ satisfies the same conditions as $G(jf)$

in the neighborhood of $f \to 0$ and $f = f_e/2$, assuming $f_e/2$ as being much larger than the frequency of slope break of $|G(jf)|$.

c) Based on $G(s)$, determine the expression $G_2(z)$ of the transmittance of the filter in the domain of the z variable according to the elements of the filter and to ω_{bl} through the modified bilinear transformation, where ω_{bl} is the multiplicative factor in this transformation (same notation as in section 1.4.6.2).

d) From $G(s)$, determine the expression $G_3(z)$ of the transmittance of this filter in the domain of the z variable in the approximation of the preserved impulse response. Compare $G_1(z)$, $G_2(z)$ and $G_3(z)$, and give a general expression that can represent the three transfer functions using four real and positive parameters, b_1, b_2, b_3 and b_4.

2) Draw the block diagram and establish the transfer function in z of the PLL by adding the correction proportional to the phase shift of question 3 of the previous exercise, as well as the other functional blocks. Give the expression of the transfer function in z of the open-loop $T(z)$ and closed-loop $H(z)$.

Answers:

1) a) We have an inverter arrangement using operational amplifiers with $Z_2 = R_2$ and $Z_1(s) = \dfrac{1}{C_1 s} + R_1$. Hence a gain $G(s) = -\dfrac{R_2\, C_1 s}{1 + R_1\, C_1 s}$, which corresponds to a phase-leading or high-pass correction filter.

The Bode diagram of $|G(jf)|$ comprises an asymptote of positive slope +1 in log-log scale or +20 dB/decade in semi-log scale, and a horizontal asymptote beyond the frequency $1/(2\pi R_1 C_1)$, with a modulus equal to R_2/R_1.

b) For $f \to 0$, we have at first-order $z^{-1} = \exp(-j2\pi f/f_e) \approx 1 - j2\pi f/f_e$. It is thus necessary that $1 - z^{-1}$ be present in the numerator to obtain a transmittance proportional to f in low frequencies. To this end, we adopt as input $E(z)=V_3(z)$ and we link $V_1(z)$ and $V_2(z)$ to $S(z)$ in the first diagram of exercise 1.6.2, which gives: $G_1(z) = \dfrac{S(z)}{E(z)} = \dfrac{a_3\,(1 - z^{-1})}{1 + a_2 - (1 + a_1) z^{-1}}$.

For $f \to 0$, $G_1(jf)\big|_{f\to 0} = \dfrac{j2\pi\, a_3 f / f_e}{a_2 - a_1 + j2\pi(1+a_1) f / f_e}$ is obtained, which gives a transmittance of the same type as $G(jf)$ but of opposite sign given that we take $a_2 > a_1$.

For $f = f_e/2$, $z^{-1} = \exp(-j\pi) = -1$, from which it is deduced that $G_1(jf_e/2) = \dfrac{2\,a_3}{2 + a_2 + a_1}$ which is real such as the high-frequency asymptote of $G(jf)$ and always positive. This justifies the previous condition that would cause a change of sign not present in $G(jf)$ if it were not satisfied. We will just take the precaution to change the reference voltage of the phase-shift detector in order to find again the plus sign of question 1 in the previous exercise.

c) As part of the modified bilinear transformation, the variable s of the transmittance $G(s)$ is replaced by $\omega_{bl}\dfrac{z-1}{z+1}$, where $\omega_{bl} = \dfrac{2\pi f_c}{\tan\left(\pi \dfrac{f_c}{f_e}\right)}$, to cancel the error in the frequency at the frequency $f_c = 1/(2\pi R_1 C_1)$, which yields:

$$G_2(z) = -\frac{R_2\, C_1 \omega_{bl} \dfrac{z-1}{z+1}}{1 + R_1\, C_1 \omega_{bl} \dfrac{z-1}{z+1}} = -\frac{R_2\, C_1 \omega_{bl}(1 - z^{-1})}{1 + R_1\, C_1 \omega_{bl} + (1 - R_1\, C_1 \omega_{bl}) z^{-1}}.$$

d) For the high-pass filter, the impulse response is given following the table of section 1.4.2 by $\delta(t) - \omega_c \mathrm{U}(t)\, \exp(-\omega_c kT_e)$ and the corresponding ZT by $\dfrac{1 - \omega_c T_e - e^{-\omega_c T_e} z^{-1}}{1 - e^{-\omega_c T_e} z^{-1}}$ where $\omega_c = 1/(R_1 C_1)$. Therefore, we directly have $G_3(z) = \dfrac{1 - \omega_c T_e - e^{-\omega_c T_e} z^{-1}}{1 - e^{-\omega_c T_e} z^{-1}}$.

It can be observed that $G_1(z)$ and $G_2(z)$ have similar expressions with respect to z, while $G_3(z)$ is different in order to satisfy the conservation condition of the impulse response. However, we can write that $G(z) = \dfrac{b_4 - b_3 z^{-1}}{b_2 - b_1 z^{-1}}$ in any case.

2) The functional blocks that have to be added are those that allow us to obtain the integral of the phase shift by means of the capacitor C_0, the previous filter, the correction proportional to the phase shift with a real transmittance d in volts/rd (which can be also viewed as a derivative correction of the integral phase shift if the left branch of the block were connected at the output of the first integrator including the capacitance C_0, as depicted in the next exercise), the assembly VCO jointly with the frequency divider with the same transmittance as previously and the transformation block of frequency into phase shift using an integrator:

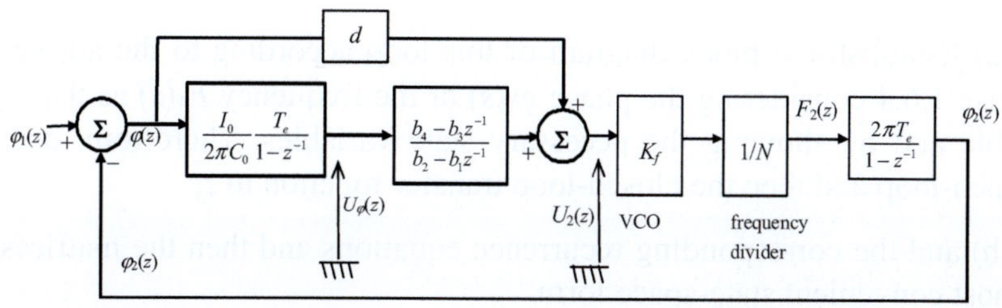

from which it can inferred:

$$T(z)=\left(\frac{I_0}{2\pi C_0}\frac{T_e}{1-z^{-1}}\frac{b_4-b_3z^{-1}}{b_2-b_1z^{-1}}+d\right)\frac{K_f}{N}\frac{2\pi T_e}{1-z^{-1}}=$$

$$\frac{K_f}{N}\left[\frac{I_0}{C_0}\frac{T_e^2}{\left(1-z^{-1}\right)^2}\frac{b_4-b_3z^{-1}}{b_2-b_1z^{-1}}+\frac{2\pi T_e d}{1-z^{-1}}\right]$$

and

$$H(z)=\frac{T(z)}{1+T(z)}=\frac{\dfrac{K_f}{N}\left[\dfrac{I_0}{C_0}\dfrac{T_e^2}{\left(1-z^{-1}\right)}\dfrac{b_4-b_3z^{-1}}{b_2-b_1z^{-1}}+2\pi T_e d\right]}{\left(1-z^{-1}\right)+\dfrac{K_f}{N}\left[\dfrac{I_0}{C_0}\dfrac{T_e^2}{\left(1-z^{-1}\right)}\dfrac{b_4-b_3z^{-1}}{b_2-b_1z^{-1}}+2\pi T_e d\right]}.$$

1.6.6. *Discrete-time systems in state-space form*

1) Second-order switched-capacitor universal filter (Figure 1.49) already studied in exercise 1.6.2.

By revisiting the diagram of the charge transfer functions established in question 4 of exercise 1.6.2, or even the two equations of conservation of zero current on each operational amplifier input in the plane of the z variable, deduce the two recurrence relations between the input voltage $E[k]$ and the two state variables $S_1[k]$, $S_1[k-1]$, $S_2[k]$, $S_2[k-1]$ at times numbered by indices k and $k-1$. Build the state matrix in the two forms of state-space representation. Determine the eigenvalues, and compare with the result of question 4 of exercise 1.6.2.

2) Double corrector PLL (question 3 of exercise 1.6.4)

a) Establish the block diagram of this loop according to the answer of exercise 1.6.4 considering the phase $\varphi_1(s)$ or the frequency $F_1(s)$ as the input variable and by showing the necessary state variables. Therefrom, deduce the open-loop and then the closed-loop transfer function in z;

b) and the corresponding recurrence equations and then the matrices of the most convenient state-space form.

3) Digital filter

By introducing the necessary state variables, establish the recurrence equations that reflect the numerical algorithm of the direct version I of Figure 1.58 by simply restricting the diagrams to the connections drawn in solid line. Deduce thereof the matrix of the most convenient state-space representation, then the poles of the transmittance and their interpretation.

Answer:

1) Second-order switched-capacitor universal filter

We directly deduce from the diagram or from the equations in z^{-1} :

$$a_1 E[k-1] + a_3 (E[k-1] - E[k]) - a_5 S_1[k] + (S_1[k-1] - S_1[k]) - a_7 S_2[k] = 0$$

and

$$a_4 (S_1[k-1] - S_1[k]) + a_2 E[k-1] - a_6 S_2[k] + (S_2[k-1] - S_2[k]) = 0 .$$

The equations are rearranged to obtain a single variable at the time numbered $k-1$ in the first member, which only requires replacing $S_1[k-1]$ in the second equation by its expression taken from the first. The result is thus the following system, according to the second form of the state-space representation:

$$\begin{cases} S_1[k-1] = (1+a_5)S_1[k] + a_7 S_2[k] + a_3 E[k] - (a_1 + a_3)E[k-1] \\ S_2[k-1] = -a_4 a_5 S_1[k] + ((1+a_6) - a_4 a_7)S_2[k] - a_3 a_4 E[k] + (a_4(a_1 + a_3) - a_2)E[k-1], \end{cases}$$

which can be written in matrix form:

$$\begin{bmatrix} S_1[k-1] \\ S_2[k-1] \end{bmatrix} = \begin{bmatrix} (1+a_5) & a_7 \\ -a_4 a_5 & ((1+a_6) - a_4 a_7) \end{bmatrix} \begin{bmatrix} S_1[k] \\ S_2[k] \end{bmatrix} + \begin{bmatrix} a_3 \\ -a_3 a_4 \end{bmatrix} E[k] + \begin{bmatrix} -(a_1 + a_3) \\ (a_4(a_1 + a_3) - a_2) \end{bmatrix} E[k-1] \cdot$$

The first matrix of the second member is the state matrix **A'** in the second notation form. Its eigenvalues are solutions of the characteristic equation

$$(1 + a_5 - \mu)((1 + a_6) - a_4 a_7 - \mu) + a_4 a_5 a_7 = 0,$$

which can be rewritten as $\mu^2 + (a_4 a_7 - 2 - a_5 - a_6)\ \mu + (1 + a_5)(1 + a_6) - a_4 a_7 = 0$

whose first member is exactly the same as the expression of the denominator of the transmittances determined in exercise 1.6.2 with μ as variable instead of z^{-1}. Thereby the roots are the same, and if they are denoted by μ_1 and μ_2, the denominator can be rewritten in the form $(z^{-1} - \mu_1)(z^{-1} - \mu_2)$.

The poles in the plane of the z variable are thus:

$$\mu_{1,2}^{-1} = \frac{2}{2 + a_5 + a_6 - a_4 a_7 \pm \sqrt{(2 + a_5 + a_6 - a_4 a_7)^2 - 4((1 + a_5)(1 + a_6) - a_4 a_7)}}$$

real if the quantity under the radical is positive and complex conjugated if it is negative.

By inverting $\mathbf{A'}$, we obtain the state matrix of the first form of the representation

$$\mathbf{A}=\frac{1}{((1+a_6)-a_4a_7)(1+a_5)+a_4a_5a_7}\begin{bmatrix}((1+a_6)-a_4a_7) & -a_7\\ a_4a_5 & (1+a_5)\end{bmatrix}.$$

2) Double corrector PLL

The block diagram of exercise 1.6.4 is transformed to take into account the modification discussed in question 3:

a) The integral phase-shift detector is an integrator with transmittance $A_1'(s)=-\frac{I_0}{2\pi C_0 s}$, which becomes $\frac{-I_0\,T_e}{2\pi C_0(1-z^{-1})}$ in the domain of the z variable according to section 1.4.1. It can be noted that $I_0\,T_e$ is the maximal charge that the capacitor can store during a sampling period and therefore $I_0\,T_e/C_0$ is its saturation voltage. The recurrence relation is deduced for the state variable U_φ according to section 1.5: $\frac{-I_0\,T_e}{2\pi C_0}\varphi[k]=U_\varphi[k]-U_\varphi[k-1]$.

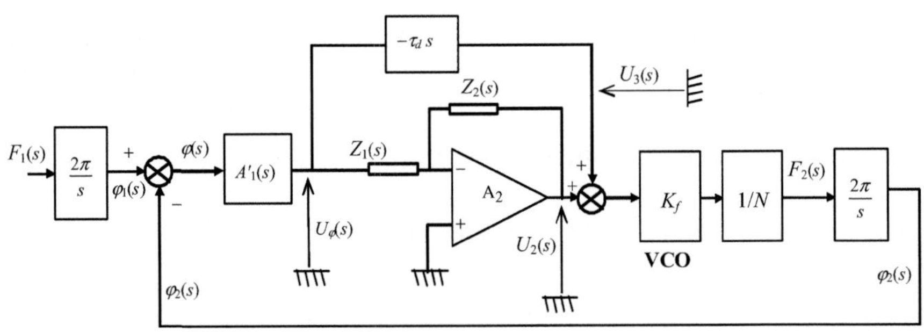

The correction filter based on operational amplifiers achieves the analog transfer function $-\left(\frac{R_2}{R_1}\frac{s}{s+\frac{1}{R_1C_1}}+\tau_d s\right)$ which becomes

$-\left(\dfrac{R_2}{R_1}\dfrac{1-\omega_1 T_e - e^{-\omega_1 T_e}z^{-1}}{1-e^{-\omega_1 T_e}z^{-1}} + \dfrac{\tau_d}{T_e}(1-z^{-1})\right)$ according to section 1.4.1 and section 1.4.2, with $\omega_1 = \dfrac{1}{R_1 C_1}$. A new state variable $U_3(z)$ is introduced so as to be able to write the first-order recurrence relations by separating the two terms of the previous transmittance:

$$U_2[k] - e^{-\omega_1 T_e} U_2[k-1] = -\frac{R_2}{R_1}\left((1-\omega_1 T_e) U_\varphi[k] - e^{-\omega_1 T_e} U_\varphi[k-1]\right)$$

and $U_3[k] = -\dfrac{\tau_d}{T_e}\left(U_\varphi[k] - U_\varphi[k-1]\right)$.

However, since $\dfrac{-I_0\, T_e}{2\pi C_0}\varphi[k] = U_\varphi[k] - U_\varphi[k-1]$, it follows that $U_3[k] = \dfrac{I_0\, \tau_d}{2\pi C_0}\varphi[k]$, which shows that U_3 and φ are not independent variables since they are related by a proportionality relation at all times.

The sum of $U_2[k]$ and $U_3[k]$ is then multiplied by K_f/N to give $F_2[k] = \dfrac{K_f}{N}\left(U_2[k] + U_3[k]\right)$ then $\varphi_2[k] - \varphi_2[k-1] = 2\pi T_e\, F_2[k]$ because $\varphi_2(z) = \dfrac{2\pi T_e}{1-z^{-1}} F_2(z)$.

The loop transmittance in z is deduced: $T_3(z) = \dfrac{I_0\, T_e}{2\pi C_0 (1-z^{-1})}\left(\dfrac{R_2}{R_1}\dfrac{1-\omega_1 T_e - e^{-\omega_1 T_e}z^{-1}}{1-e^{-\omega_1 T_e}z^{-1}} + \dfrac{\tau_d}{T_e}(1-z^{-1})\right)\dfrac{K_f}{N}\dfrac{2\pi T_e}{1-z^{-1}}$,

let $T_3(z) = \dfrac{I_0\, T_e^2 K_f}{C_0\, N}\dfrac{R_2}{R_1}\dfrac{1-\omega_1 T_e - e^{-\omega_1 T_e}z^{-1}}{\left(1-z^{-1}\right)^2\left(1-e^{-\omega_1 T_e}z^{-1}\right)} + \dfrac{I_0\, T_e\, \tau_d\, K_f}{C_0\, N}\dfrac{1}{1-z^{-1}}$ and the closed-loop transmittance is $H_3(z) = \dfrac{\varphi_2(z)}{\varphi_1(z)} = \dfrac{T_3(z)}{1+T_3(z)}$. Since we are looking for the transmittance $\frac{F_2(z)}{F_1(z)}$ where we have both $F_1(z) = \frac{1-z^{-1}}{2\pi T_e}\varphi_1(z)$ and $F_2(z) = \frac{1-z^{-1}}{2\pi T_e}\varphi_2(z)$, we also have: $\frac{F_2(z)}{F_1(z)} = H_3(z)$.

b) The synthesis of the recurrence equations is carried out by employing the second form of the state-space representation and by replacing $U_\varphi[k-1]$ by the second member of the first relation in the following:

$$U_\varphi[k-1] = U_\varphi[k] + \frac{I_0\,T_e}{2\pi C_0}\varphi[k] \tag{1}$$

$$U_2[k-1] = e^{\omega_1 T_e}U_2[k] + \frac{R_2}{R_1}e^{\omega_1 T_e}(1-\omega_1 T_e)U_\varphi[k] - \frac{R_2}{R_1}\left(U_\varphi[k] + \frac{I_0\,T_e}{2\pi C_0}\varphi[k]\right) \tag{2}$$

Then by shifting from k to $k-1$, relation (3) is written:

$$F_2[k-1] = \frac{K_f}{N}\left[e^{\omega_1 T_e}U_2[k] + \frac{R_2}{R_1}e^{\omega_1 T_e}(1-\omega_1 T_e)U_\varphi[k]\right]$$
$$-\frac{K_f}{N}\left[\frac{R_2}{R_1}\left(U_\varphi[k] + \frac{I_0\,T_e}{2\pi C_0}\varphi[k]\right) - \frac{I_0\,\tau_d}{2\pi C_0}\varphi[k-1]\right]$$

where $\varphi[k-1]$ will be replaced according to the state variables at time k at the end of the procedure and $\varphi_2[k-1] = \varphi_2[k] - 2\pi T_e\,F_2[k]$ (4).

Finally, it is necessary to introduce the feedback loop and the subtractor, with $\varphi[k] = \varphi_1[k] - \varphi_2[k]$ in which $\varphi_1[k]$ is an input variable. By shifting from k to $k-1$, the last equation is written as follows:

$$\varphi[k-1] = \varphi_1[k-1] - \varphi_2[k] + 2\pi T_e\,F_2[k] \tag{5}$$

We thus have in total five state variables: U_φ, U_2, F_2, φ_2 and φ.

The input variable φ_1 depends on the other possible input variable F_1 according to $F_1[k] = \dfrac{1}{2\pi T_e}(\varphi_1[k] - \varphi_1[k-1])$.

By replacing $\varphi[k-1]$ in relation (3) by its expression taken from relation (5), the system can be written in matrix form and we can establish the expression of the state matrix **A'**:

$$\begin{bmatrix} U_\varphi[k-1] \\ U_2[k-1] \\ F_2[k-1] \\ \varphi_2[k-1] \\ \varphi[k-1] \end{bmatrix} = \mathbf{A'} \begin{bmatrix} U_\varphi[k] \\ U_2[k] \\ F_2[k] \\ \varphi_2[k] \\ \varphi[k] \end{bmatrix} + \mathbf{B'} \qquad \text{with:}$$

$$\mathbf{A'} = \begin{bmatrix} 1 & 0 & 0 & 0 & \dfrac{I_0 T_e}{2\pi C_0} \\ \dfrac{R_2}{R_1}\left(e^{\omega_1 T_e}(1-\omega_1 T_e)-1\right) & e^{\omega_1 T_e} & 0 & 0 & -\dfrac{R_2}{R_1}\dfrac{I_0 T_e}{2\pi C_0} \\ \dfrac{K_f R_2}{N R_1}\left(e^{\omega_1 T_e}(1-\omega_1 T_e)-1\right) & \dfrac{K_f}{N}e^{\omega_1 T_e} & \dfrac{K_f}{N}\dfrac{I_0 \tau_d T_e}{C_0} & -\dfrac{K_f}{N}\dfrac{I_0 \tau_d}{2\pi C_0} & -\dfrac{K_f}{N}\dfrac{R_2 I_0 T_e}{2\pi R_1 C_0} \\ 0 & 0 & -2\pi T_e & 1 & 0 \\ 0 & 0 & 2\pi T_e & -1 & 0 \end{bmatrix}$$

$$\text{and } \mathbf{B'} = \mathbf{B}_1'[\varphi_1[k-1]] = \begin{bmatrix} 0 \\ 0 \\ \dfrac{K_f}{N}\dfrac{I_0 \tau_d}{2\pi C_0} \\ 0 \\ 1 \end{bmatrix} [\varphi_1[k-1]]$$

If F_1 is the input variable, $\varphi_1[k-1]$ will be deduced from $\varphi_1[k-1] = \varphi_1[k] - 2\pi T_e F_1[k]$.

$$\text{In this case, } \mathbf{B'} = \mathbf{B}_0' \begin{bmatrix} F_1[k] \\ \varphi_1[k] \end{bmatrix} = \begin{bmatrix} 0 & 0 \\ 0 & 0 \\ -\dfrac{K_f}{N}\dfrac{I_0 T_e \tau_d}{C_0} & \dfrac{K_f}{N}\dfrac{I_0 \tau_d}{2\pi C_0} \\ 0 & 0 \\ -2\pi T_e & 1 \end{bmatrix} \begin{bmatrix} F_1[k] \\ \varphi_1[k] \end{bmatrix}.$$

To obtain the first form of the state-space representation that makes it possible to use recurrence computation of all samples from $k = 1$, the series of input samples being given, it is necessary to invert matrix **A'** and to rewrite the first matrix equation taking into account the relations indicated in section 1.5. This type of computation can be done using numerical computation software (MATLAB, SciLab, etc.), which requires the prior scaling of loop parameters, as in exercise 1.6.4. In principle, it is possible to take into account the effect of modulation or frequency jumps for the sampling period T_e to the extent that it appears as a parameter in many matrix elements. It is nonetheless important to properly analyze the relationship between frequencies F_1, F_2 and $1/T_e$ as it is represented in Figure 1.26 but with different frequencies F_1 and F_2, and therefore an evolutive interval Δt, which will itself induce the update of φ at every sampling step.

If F_2 and φ_2 are output variables, we will write the second matrix equation: $\begin{bmatrix} F_2[k] \\ \varphi_2[k] \end{bmatrix} = \begin{bmatrix} 0 & 0 & 0 & 1 & 0 & 0 \\ 0 & 0 & 0 & 0 & 1 & 0 \end{bmatrix} \begin{bmatrix} U_\varphi[k] \\ U_2[k] \\ U_3[k] \\ F_2[k] \\ \varphi_2[k] \\ \varphi[k] \end{bmatrix}$, which is however not essential since the output variables are part of the state variables.

3) Digital filter

The following variables are introduced in the diagram:

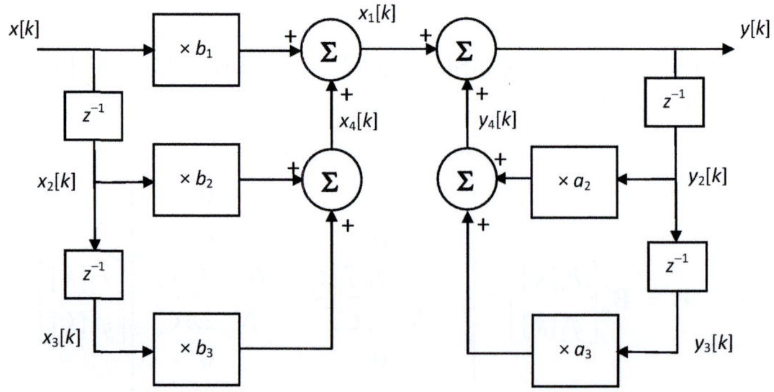

We then have: $x_2[k] = x[k-1]$; $x_3[k] = x_2[k-1]$; $x_4[k] = b_2\, x_2[k] + b_3\, x_3[k]$; $x_1[k] = x_4[k] + b_1\, x[k]$;

and $y_2[k] = y[k-1]$; $y_3[k] = y_2[k-1]$; $y_4[k] = a_2\, y_2[k] + a_3\, y_3[k]$; $y[k] = y_4[k] + x_1[k]$.

The necessary carryovers are made such that the summation relations include variables both at times $k-1$ and k: $x_4[k] = b_2\, x_2[k] + b_3\, x_2[k-1]$; $y_4[k] = a_2\, y_2[k] + a_3\, y_2[k-1]$; and then $x_1[k] = b_2\, x_2[k] + b_3\, x_2[k-1] + b_1\, x[k]$; $y[k] = a_2\, y_2[k] + a_3\, y_2[k-1] + x_1[k]$, that is rewritten as:

$x_1[k]= b_2\, x[k-1] + b_3\, x_2[k-1] + b_1\, x[k]$;

then: $y[k]= a_2\, y[k-1] + a_3\, y_2[k-1] + b_2\, x[k-1] + b_3\, x_2[k-1] + b_1\, x[k]$ where $x[k]$ is the input variable. All the recurrence relations are then available which allows us to directly write the state-space representation of the first form, using the state variables x_2, x_3, y_2 and y.

$$\begin{bmatrix} x_2[k] \\ x_3[k] \\ y_2[k] \\ y[k] \end{bmatrix} = \begin{bmatrix} 0 & 0 & 0 & 0 \\ 1 & 0 & 0 & 0 \\ 0 & 0 & 0 & 1 \\ b_3 & 0 & a_3 & a_2 \end{bmatrix} \begin{bmatrix} x_2[k-1] \\ x_3[k-1] \\ y_2[k-1] \\ y[k-1] \end{bmatrix} + \begin{bmatrix} 0 \\ 0 \\ 0 \\ b_1 \end{bmatrix} \left[x[k]\right] + \begin{bmatrix} 1 \\ 0 \\ 0 \\ b_2 \end{bmatrix} \left[x[k-1]\right] \cdot$$

The eigenvalues of the state matrix are the solution of

$$\det \begin{bmatrix} -\lambda & 0 & 0 & 0 \\ 1 & -\lambda & 0 & 0 \\ 0 & 0 & -\lambda & 1 \\ b_3 & 0 & a_3 & a_2-\lambda \end{bmatrix} = 0 .$$

By expanding this determinant with respect to the first element, the minor obtained by removing the first row and the first column reads as follows:

$-\lambda\left(\lambda^2(a_2-\lambda)+a_3\lambda\right)=0$, which gives two zero eigenvalues $\lambda_{1,2}=0$, and the roots of $\lambda^2 - a_2\lambda - a_3 = 0$, that is $\lambda_{3,4} = \dfrac{1}{2}\left(a_2 \pm \sqrt{a_2^2 - 4a_3}\right)$, which are thus the transmittance poles in the plane of the z variable. The two poles at the origin correspond to a factor z^2 in the denominator of the transmittance, that is, a factor z^{-2} in the numerator, which is equivalent to a delay of two sampling periods in the discrete-time domain and which indicates that the result of the computation performed by the filter is obtained after this delay. In the expression of the poles which determine the denominator of the transmittance, only the multiplicative coefficients a_2 and a_3 which are included in the right-hand side of the diagram, are present. This is logical insofar as it is the only part of the diagram where recursivity is present due to the operations carried out on the output samples $y[k]$. There are two real poles in the plane of the variable z if $a_2^2 > 4a_3$, and two complex conjugated poles otherwise. These two poles induce a second-order behaviour. It is also logical that coefficients b_1, b_2 and b_3 play no role in the pole expression of $\lambda_{3,4}$ because they modify the input samples without recursivity, inducing a linear combination of the input samples $x[k]$, $x[k-1]$ and $x[k-2]$ sent to the right part of the filter. For this reason, b_1, b_2 and b_3 will appear only in the numerator of the transmittance, which does not influence the poles in any way.

2

Quantized Level Systems: Digital-to Analog and Analog-to-Digital Conversions

2.1. Quantization noise

Systems that establish a relationship between a signal represented by a continuous function or an *analog signal* x, and a signal represented by an *integer* M or quantized signal, perform approximations that can be represented by a multistep function (Figure 2.1):

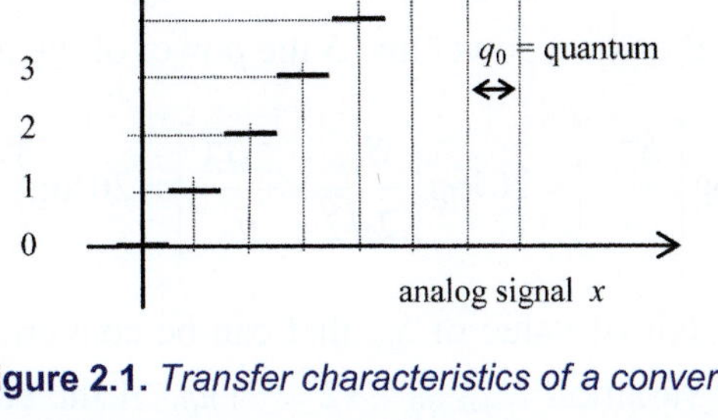

Figure 2.1. *Transfer characteristics of a converter*

The largest variation possible of x which leaves M invariant is the quantum q_0; there is then a maximal error of quantization $\pm\frac{q_0}{2}$ during conversion: $x = M q_0 \pm \varepsilon \frac{q_0}{2}$, where ε is a random number comprised between 0 and 1.

<u>Up to the quantization error, the system is thus correctly a linear system.</u>

With rare exceptions, the natural binary code will be used to encode the positive number M (as in Figure 2.1).

Since there is no prior information available about the value of x, the probability of the interval Δx between x and Nq_0 is uniformly distributed over the interval q_0. The variance (proportional to power) is then equal to, $\frac{1}{q_0}\int_{-q_0/2}^{q_0/2} \Delta x^2 \, dx = \frac{1}{q_0}\left[\frac{\Delta x^3}{3}\right]_{-q_0/2}^{q_0/2} = \frac{q_0^2}{12}$, which corresponds to a standard deviation $\frac{q_0}{2\sqrt{3}}$, called RMS (root mean square) quantization noise.

The comparison with the standard deviation of the signal allows the signal-to-noise ratio to be evaluated. In the case of a periodic deterministic signal of period T, the standard deviation is comparable to the RMS value S_{eff} of its alternative part, because in the expression $S_{eff}^2 = \frac{1}{T}\int_0^T \left[s(t) - S_{avg}\right]^2 dt$, dt/T represents the probability that the signal be comprised between $s(t)$ and $s(t) + ds$. In the case of a sinusoidal signal of peak-to-peak value S_{pp}, we have $S_{eff} = \frac{S_{pp}}{2\sqrt{2}}$. The signal-to-noise ratio can thus be calculated in decibel, where N represents the noise power and S the power of the useful signal:

$$(S/N)_{\text{dB}} = 10\log\left[\frac{S_{eff}^2}{N}\right] = 20\log\left[\frac{S_{pp}}{2\sqrt{2}} \times \frac{2\sqrt{3}}{q_0}\right] = 20\log\left[\frac{S_{pp}}{q_0}\right] + 1.76\text{dB}.$$

However, the maximal value of S_{pp} that can be converted is equal to the so-called full-scale variation $N_{max}\, q_0 = (2^n - 1)q_0$ if the encoding is binary and achieved onto n bits. In the case where $2^n >> 1$, the maximal signal-over-noise ratio can be established:

$$(S/N)_{\text{dB},max} = 20\log\left(2^n\right) + 1.76\text{dB} = 6.02n + 1.76\text{dB}.$$

Another aspect of the conversion relates to the distribution of the noise power in the frequency domain, resulting from the time rate of this conversion that can be characterized by a time period T_e corresponding to the time required to perform the conversion. In most cases, it can be considered that this is a constant sampling period. However, there is no correlation between the error $\Delta x(t)$ and the errors $\Delta x(t+T_e)$ or $\Delta x(t-T_e)$ in previous or following sampling periods. The autocorrelation function $N_{xx}(\tau)=\int_{-\infty}^{+\infty}\Delta x(t)\Delta x(t-\tau)dt$ of Δx is therefore a non-zero constant only for a delay τ not exceeding the period T_e since then the probability that Δx remains the same during this period is equal to 1. Since the power spectrum of the noise is the FT of $N_{xx}(\tau)$ (see Chapter 1, Volume 2), [MUR 17b] we obtain the noise power density by

$$D_N(f)=\frac{q_0^2}{12}\int_{-T_e/2}^{+T_e/2}\exp(-j2\pi f\tau)d\tau=\frac{q_0^2}{12}T_e\frac{\sin(\pi fT_e)}{\pi fT_e}=\frac{q_0^2}{12f_e}\frac{\sin(\pi fT_e)}{\pi fT_e}$$

taking into account the variance[1]. The power of the quantization noise is therefore significant up to the frequencies of the order of the sampling frequency f_e or the inverse of the time needed to complete the conversion, with a spectral density equal to $\frac{q_0^2}{12f_e}$ in low frequency, expressed in V^2/Hz if q_0 is a voltage or A^2/Hz if q_0 is a current.

A commonly adopted approximation consists of considering a noise power density that is constant and equal to $\frac{q_0^2}{12f_e}$ in the interval $[-f_e/2, f_e/2]$ and zero outside because $f_e/2$ is close to the -3 dB cut-off frequency of $D_N(f)$.

---

1 Indeed, given that $\int_{-\infty}^{+\infty}\frac{\sin x}{x}dx=\pi$ as it can be proven by taking the integral of $\exp(jx)/x$ following a contour in the complex plane, $\int_{-\infty}^{+\infty}\frac{\sin(\pi fT_e)}{\pi fT_e}df=f_e$ by making the change of variable $x=\pi fT_e$, which implies that $N=\int_{-\infty}^{+\infty}D_N(f)df=q_0^2/12$.

2.2. Characteristics of converters

The two types of conversion are performed either by means of digital-to-analog converters (DACs), or analog-to-digital converters (ADCs), symbolized in Figure 2.2:

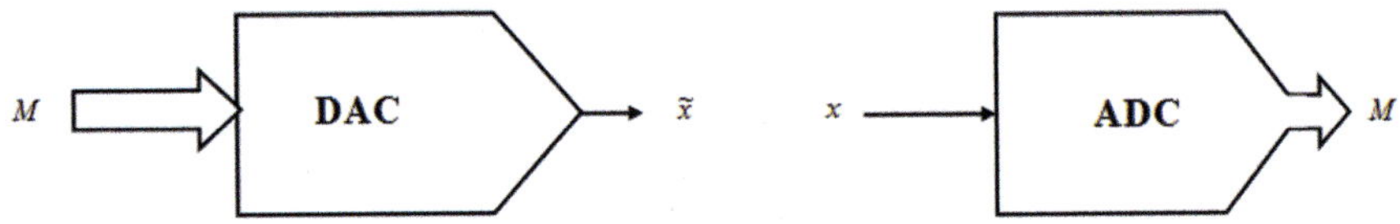

Figure 2.2. *Possible symbols for converters*

Regarding DACs, the output signal $\tilde{x}$ is pseudo-analog because it consists of discrete values of current or voltage. The number M is subsequently referred to as either the "code" or digital data.

2.2.1. *Dynamics and resolution*

From the moment that $2^n >> 1$, the dynamics can be likened to the maximal signal-to-noise ratio assessed in section 2.1.

The resolution is generally defined as being the smallest variation possible in relation to the full scale; in other words $\frac{1}{2^n}$; however resolution is commonly employed to designate the number of bits n. In binary coding, the quantum q_0 is often confused with LSB ("Least Significant Bit") and the maximal quantization error $\pm\frac{q_0}{2}$ is then referred to as $\pm\frac{1}{2}\text{LSB}$.

2.2.2. *Static errors*

It is easier to define static performance by reasoning from the continuous transfer function obtained by joining the midpoints of the line segments of the multistep transfer characteristic (Figure 2.3).

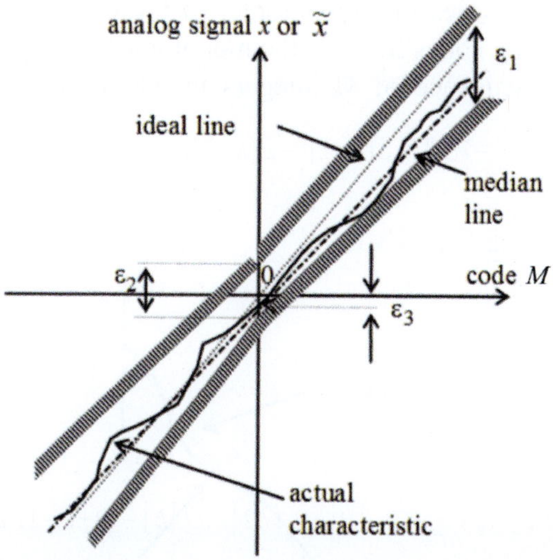

Figure 2.3. *Static errors in the actual transfer characteristic*

Absolute static performances are those of the converter that have not been subjected to any adjustment:

– Precision or full-scale error (ε_1) and zero absolute error (ε_2) define the area where the continuous transfer function must necessarily be encountered (between shaded areas in Figure 2.3).

– Offset error ε_3 is the deviation between the median line and the origin.

– Gain error is the deviation between the average slope of the continuous transfer function and that of the ideal line.

These last two errors can be adjusted to 0 but sustain temperature drifts. Assuming these last two errors to be cancelled, the relative static performances can be defined (Figure 2.4):

– Linearity error or integral linearity error ε_4 is the maximum deviation between the actual curve of the continuous transfer function and the ideal line (measured in LSB, in general).

– Monotonicity: this is measured by the differential linearity error, or more precisely the maximum deviation with respect to "1 LSB" for two adjacent codes. When this error becomes greater than 1 LSB, the continuous transfer function becomes non-monotonic. In the case of ADCs, it is said that there are missing codes because the monotonic progression of the signal x gives rise to a sequence of M integers in which adjacent numbers are missing.

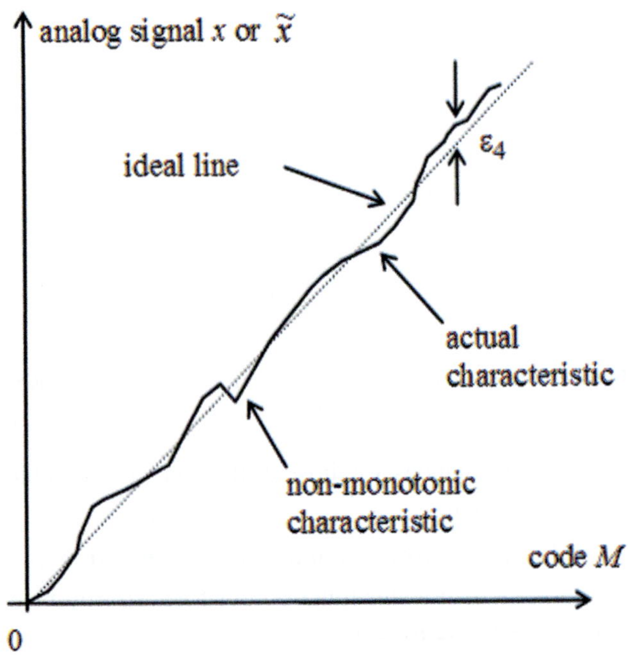

Figure 2.4. *Relative errors in the actual transfer characteristic*

2.2.3. *Dynamic operation*

In DACs, the settling time t_{sl} (Figure 2.5) is defined as the time lag between the application of a new code and the moment at which the analog output assumes its final value up to $\pm\varepsilon_5$ (according to the resolution, $\varepsilon_5 = 0.1\%$ or 0.01% or more generally $\varepsilon_5 = \frac{q_0}{2}$, namely half an LSB).

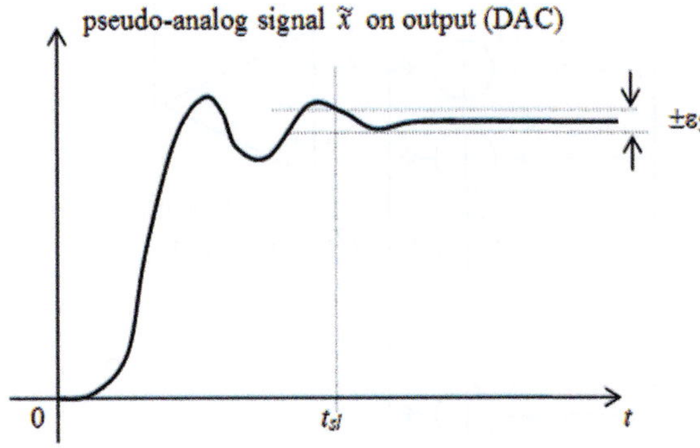

Figure 2.5. *Step response of a DAC*

For ADCs, the conversion time is defined as the time lag between the application of a unit step on the analog input and the moment when the digital output is updated.

The variation in M may also generate impulse transient states ("glitches") which translate into spikes of current or voltage when M switches between two values, especially when the bits having highest weights change. All converters are not prone however to such a flaw, depending on their type.

2.3. Digital-to-analog conversion

2.3.1. *Current- or voltage-weighted systems of 2^n dynamics in binary code*

The summation of currents or voltages proportional to the weight of each bit is performed to convert a number encoded in natural binary and expressed by $M = \sum_{k=0}^{n-1} b_k 2^k$ (Figure 2.6). In integrated circuits, since currents are proportional to the active surfaces of transistors, dynamics up to 2^{10} can be reached, most often by combining identical sources arranged in parallel with sources directly providing the desired current. On the other hand, with resistances, it is very difficult to maintain an acceptable accuracy for those that have the highest weights and as such, technology based on resistances is rarely used.

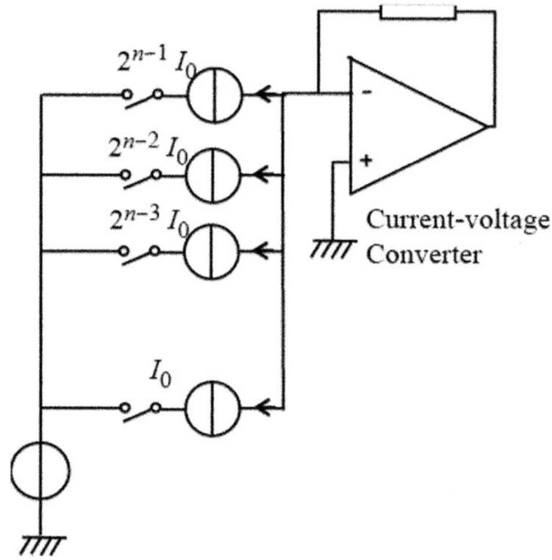

Figure 2.6. *Current-weighted converter*

2.3.2. *Iterative resistance of a network of voltage and current dividers*

Each divider performs a voltage division (Figure 2.7): $\frac{V_{k2}}{V_{k1}} = \frac{R_u R_{k2}}{R_{k1}(R_u + R_{k2}) + R_u R_{k2}} = \alpha_k$ and a current division $\frac{I_{k2}}{I_{k1}} = \frac{R_{k2}}{R_u + R_{k2}} = \beta_k$. It is imposed that R_u is the iterative resistance: $R_u = R_{k1} + \frac{R_u R_{k2}}{R_u + R_{k2}}$. By deferring, we then get $\alpha_k = \beta_k$, and therefore the same voltage and current division ratio and $R_{k1} = (1 - \alpha_k) R_u$; $R_{k2} = R_u \frac{\alpha_k}{(1 - \alpha_k)}$

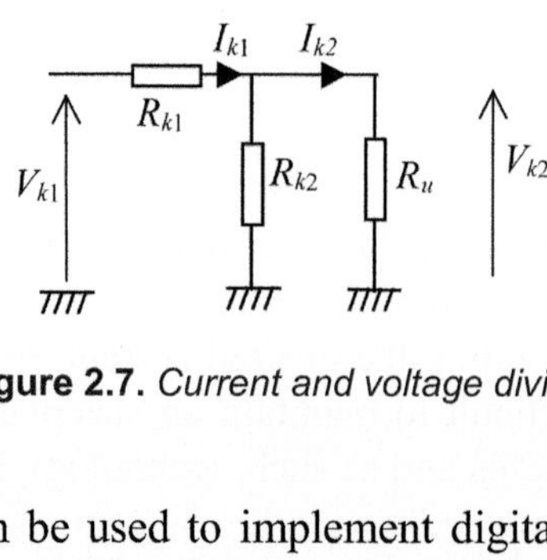

Figure 2.7. *Current and voltage divider*

Such a network can be used to implement digital attenuators controlled by logical variables, capable, for instance, of providing an attenuation between 0 and X dB in steps of 0.5 or 0.375 dB. In this case, there is no proportionality between a number M encoded in binary and resistors values, α_k and β_k being dependent on k. On the other hand, it is possible to scale resistors to obtain a logarithmic relation between M and the attenuation. This type of attenuator can constitute the nonlinear function necessary to implement a logarithmic compressor.

2.3.3. *R-2R ladders*

In the case of DACs, which make use of the n-bit binary numeral system, each divisor performs a division by 2:

$$\alpha_k = 1 - \alpha_k = \frac{1}{2} \quad \forall k \Rightarrow R_{k1} = \frac{1}{2} R_u = R \quad ; \quad R_{k2} = R_u = 2R \quad \forall k .$$

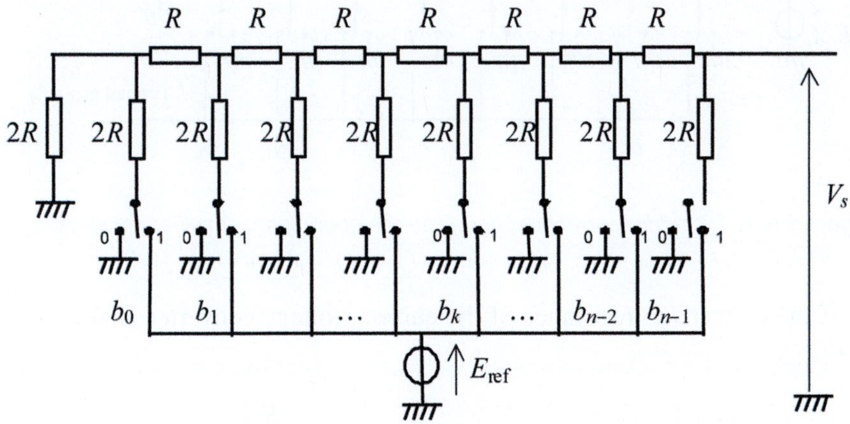

Figure 2.8. *R-2R ladder voltage divider*

R-2R ladders or R-2R division networks are deduced thereof; either voltage (Figure 2.8), or current dividers (Figure 2.9) with $b_k = 0$ or 1 if the switch is respectively on position 0 (to ground) or 1 (either to the reference source or current output).

The current division network (Figure 2.9) is preferable because the switches, which are MOSFET transistors, are still subject to the same voltage due to the virtual ground imposed by an operational amplifier (not shown) assembled as a current-voltage converter. This arrangement avoids transient charge and the discharge of parasitic capacitances. The current-voltage converter placed on output of current I_s is a means to obtain a voltage proportional to the number M encoded in natural binary, or more specifically for n-bit: $M = \sum_{k=0}^{n-1} b_k 2^k$ with $M_{max} = 2^n - 1$ (see exercise 2.6.1).

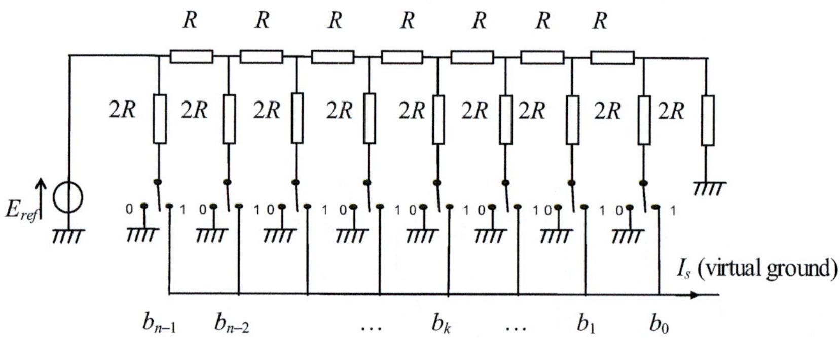

Figure 2.9. *R-2R ladder current division (with the operational amplifier not shown)*

If the conversion resistance of the current-voltage converter is also equal to R:

$$V_s = \frac{E_{ref}}{2^n} \sum_{k=0}^{n-1} b_k 2^k \text{ and quantum } q_0 \text{ is equal to } \frac{E_{ref}}{2^n}.$$

In order to obtain a symmetrical bipolar output voltage between $-E_{ref}/2$ and $+E_{ref}/2$, it is enough to add a fixed analog voltage $-E_{ref}/2$ to the current-voltage converter. If the encoding remains unchanged, this system then corresponds to a signed number M, the encoded rounding up to half-scale 2^{n-1} (see Chapter 3 of Volume 1), [MUR 17a]. The sign bit is the most significant bit (MSB), but unlike encodings used in calculators, 0 corresponds to the minus sign and 1 to the plus sign. The encoding is exactly the same as natural binary code but the number 000... 000 is put in correspondence with the number $M_{min} = -2^{n-1}$ and with the minimal voltage, that is $-E_{ref}/2$, whereas $M_{max} = 2^{n-1} - 1$. Examples with $n = 8$ bits :

Natural decimal	Binary equivalent (or binary shifted by 01111111)	Equivalent signed decimal, shifted by subtracting 127
221	11011101	94
128	10000000	1
127	01111111	0
126	01111110	–1
98	01100010	– 29

The addition of two binary numbers shifted this way only yields a number belonging to the same numbering system if we subtract thereof the decimal equivalent of 254 in an 8-bit system or more generally 111... 1110, after having taken into account the possible carryover, resulting in an extra bit of stronger weight than the original MSB. In the case where arithmetic operations must be performed following conversion, this particularity ought to be considered or otherwise dealt with using unsigned numbers.

2.3.4. *Charge redistribution capacitive converters*

In these converters, the information contained in the M number coded by its binary equivalent $M = \sum_{k=0}^{n-1} b_k 2^k$ is transcribed as the amount of charge within capacitors, which then results in obtaining a voltage proportional to M in one of them. The conversion is carried out either according to a "series" principle by which the final value is obtained after n operations successively involving every bit starting with the LSB, or a "parallel" principle by which all bits are converted at the same time, or a mixed principle. In all cases, the technology used is of the CMOS type and the redistribution of charge into the various capacitors is achieved according to an algorithm capable of controlling the CMOS switches that operate the charge transfers. These converters have the advantage of correcting the offset errors during the process of conversion itself due to the fact that they employ switched capacitors capable of memorizing them.

2.3.4.1. *Series DACs*

The basic assembly is made up of two capacitors C_0 and C_1 with the same capacitance C (Figure 2.10). C_0 is charged when submitted to a certain voltage V_0 when it is isolated from C_1 which is charged with the voltage V_1. Then in a second phase, they are put in parallel and isolated from the voltage generator V_0, which has the effect of equalizing the charges and the common voltage V'_1.

The conservation of the total charge $Q_0 + Q_1$ in the isolated system implies that $CV_0 + CV_1 = 2\,C\,V'_1$, that is also $V'_1 = (V_0 + V_1)/2$. The sequence formed by these two operations therefore performs both the summation of voltages initially at the terminals of C_0 and C_1, and the division by two of this sum.

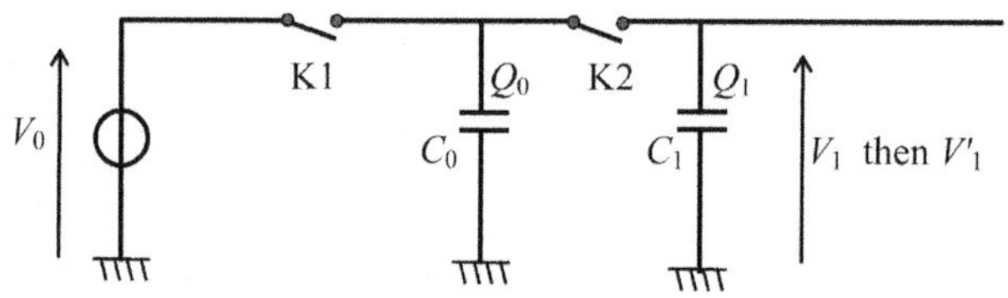

Figure 2.10. *Basic circuit for charge redistribution conversion*

If this sequence is repeated n times by introducing each time a voltage V_k ($k = 0$ to $n-1$) instead of V_0 at the terminals of C_0, the sum of all the $V_k / 2^{n-k}$ can be found at the terminals of C_1 because the voltages V_k are divided by 2 when introduced and then again for each subsequent sequence.

Finally, if $V_k = b_k\,V_{ref}$ and if b_0 corresponds to the LSB, the system will be able to supply after n sequences a voltage into the capacitor C_1 equal to $\sum_{k=0}^{n-1} b_k \frac{V_{ref}}{2^{n-k}} = \frac{V_{ref}}{2^n} \sum_{k=0}^{n-1} b_k 2^k = M \frac{V_{ref}}{2^n}$.

An initialization phase intended to cancel off the charges in the capacitors must be added beforehand (see the details of the algorithm in exercise 2.6.2, part 1). Nonetheless, the branches that do not have any point connected to the ground introduce parasitic capacities whose effect can be neutralized providing that an operational amplifier be used.

In the improved converter (Figure 2.11), an operational amplifier in a loop arrangement makes it possible to get a virtual ground apart from the offset voltage V_{off}. Therefore, capacitors have an armature that always remains at the same potential, very close to 0, which cancels out the effect of parasitic capacitances connected at the same point. The procedure includes an initialization phase that enables at time t_{-1}, the charge of capacitors to be canceled off except that due to the offset voltage of the operational amplifier, then the repetition of sequences comprising two phases respectively at times $t_{k-1/2}$ and t_k (see details in exercise 2.6.2, part 2) starting with the LSB. During the second phase, the two capacitors are placed in parallel resulting in their armatures connected to the minus input of the operational amplifier forming an isolated system. Voltages $b_0 V_{ref}$, $b_1 V_{ref}$, $b_2 V_{ref}$, ... $b_k V_{ref}$, ... $b_{n-1} V_{ref}$ are, respectively, applied to the first capacitor C_1 at times $t_{-1/2}$, $t_{1/2}$, $t_{3/2}$, ..., $t_{k-1/2}$, ... $t_{n-3/2}$. After the initialization stage, which also operates to memorize the offset voltage V_{off} of the operational amplifier, a logic function controlled by a counter provides the logic control for opening the switches if the logical variable is equal to 0, or for closing the switches if the logical variable is equal to 1 (in Figure 2.11, the + sign indicates the logical function OR and the dot the logical function AND), according to an algorithm whose result is (see exercise 2.6.2, part 2 for details)

$$Q_2(t_k) = C\left[V_{off} - \frac{V_{ref}}{2}\left(\frac{b_0}{2^k} + \frac{b_1}{2^{k-1}} + \cdots + b_k\right)\right] \text{ and}$$

$$V_s(t_k) = \frac{V_{ref}}{2}\left(\frac{b_0}{2^k} + \frac{b_1}{2^{k-1}} + \cdots + b_k\right).$$

For $k = n-1$, the same result as for the basic converter is found again, independently of the offset voltage of the operational amplifier, which is then compensated for.

The settling time needed to obtain the completely converted voltage $V_s(t_{n-1})$ therefore contains $n+1$ sequences including the initialization. Each sequence is composed of two phases which require a transfer of charge in one or two capacitors, either from the voltage generator V_{ref}, or by means of the CMOS switches. Considering that a time constant τ controls such charge transfers, they have to be over at most up to $\pm q_0/2$ (or improperly $\pm 1/2$ LSB) as a relative value so as not to introduce error. For a typical value for $n = 10$ bits, one should allow for a period $t_{k+1} - t_k$ of about 10 time constants τ, such that each charge transfer will last at least 4τ, which is enough to achieve the final state within 0.1% in a first order system, compatible with the resolution of $2^{-n} = 1/1{,}024$. By considering an order of magnitude of the time constant $\tau = 50\ \Omega \times 50$ pF $= 2.5$ ns, and a settling time of 110 times this value, we get 275 ns, which can be considered as a typical value with the chosen capacitance.

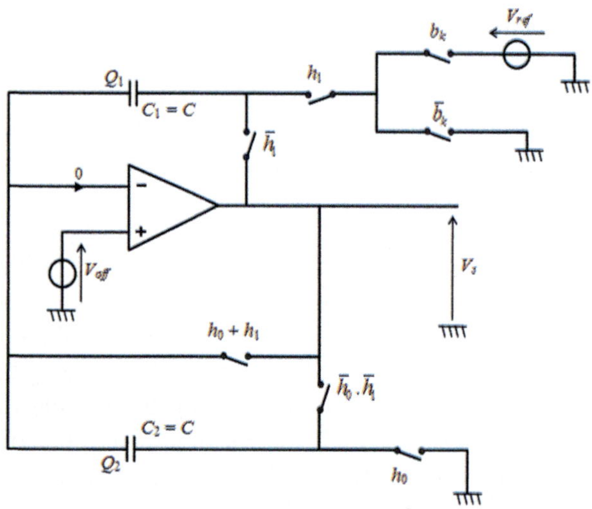

Figure 2.11. *Series converter using switched capacitors and operational amplifier*

2.3.4.2. *Parallel DACs*

Similar to the case of the current weighting-based converter, by employing n capacitors whose capacitance values are scaled within a range from C to 2^{n-1} C, it is possible to simultaneously convert all bits.

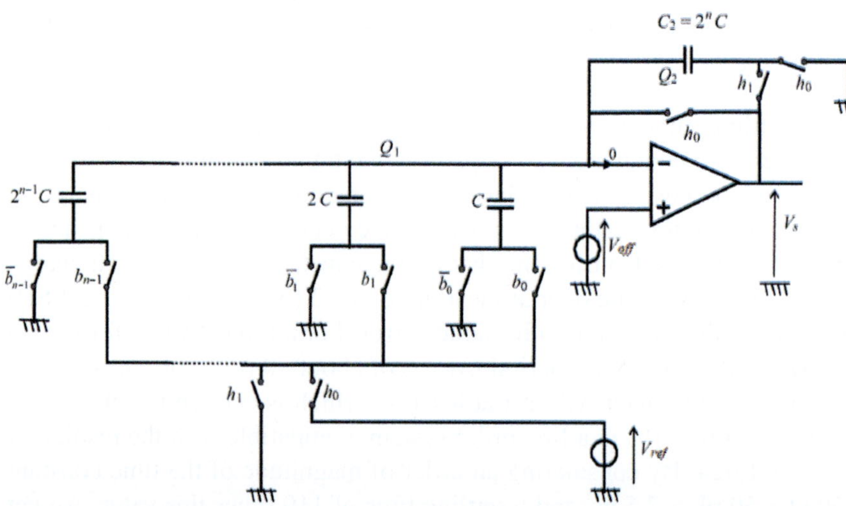

Figure 2.12. *Parallel switched-capacitor converter*

The operation comprises two phases at times t_0 and t_1. The first ($h_0 = 1$ for which switches are turned on with a closed contact, $h_1 = 0$ for which switches are turned off with an open contact) is capable of storing the offset voltage V_{off} in the capacitance C_2 as well as the charge resulting from the bits to be converted in the other capacitances: $Q_1(t_0) = CV_{off}\sum_{k=0}^{n-1} 2^k - CV_{ref}\sum_{k=0}^{n-1} b_k 2^k$. Since the operational amplifier is inserted inside a voltage follower circuit, $Q_2(t_0) = C_2 V_{off} = 2^n C V_{off}$. During the second phase ($h_0 = 0$, $h_1 = 1$), the operational amplifier switches to an amplifier circuit with C_2 in the feedback loop. Furthermore, the capacitors that contained charge Q_1 are all connected to the ground through the switches since either $\bar{b}_k$, or b_k is equal to 1, which implies a conductive path to the ground in both cases because the switch h_1 is closed. The redistribution of charge is achieved through conservation of charge in the isolated system comprising the armatures of all the capacitors connected to the minus input of the operational amplifier:

$Q_1(t_0) + Q_2(t_0) = Q_1(t_1) + Q_2(t_1)$, where the second member is determined by voltages V_{off} and V_s. Expressing each term as a function of the stored charge gives:

$$CV_{off}\sum_{k=0}^{n-1} 2^k - CV_{\text{ref}}\sum_{k=0}^{n-1} b_k 2^k + 2^n CV_{\text{off}} = CV_{\text{off}}\sum_{k=0}^{n-1} 2^k + 2^n C\left(V_{\text{off}} - V_s\right),$$

whose result provides the voltage converted into a single sequence $V_s = \frac{V_{\text{ref}}}{2^n}\sum_{k=0}^{n-1} b_k 2^k = M\frac{V_{\text{ref}}}{2^n}$.

The advantage of this system can be appreciated in its speed which is nonetheless hampered by the total capacitance equal to 2^{n+1} C, because it becomes all the more significant that the number of bits is higher. Consequently, the time constant upon which the redistribution of charge depends will be much greater than in the series converter and the resulting speed gain will be decreased. In addition, conservation of sufficient accuracy for the largest capacitances will become even more problematic when the number of bits will be greater.

To reduce the previous disadvantages, we can build a system that decreases the ratio of extreme capacitances in large proportions, namely $2^{n/2}$ instead of 2^n, by splitting the group of capacitors into two sections. Half of the bits of lowest weight control the capacitors' charge of part A of the assembly, while half of the bits of strongest weights control the charge of capacitors in part B, having values identical to those of part A (Figure 2.13).

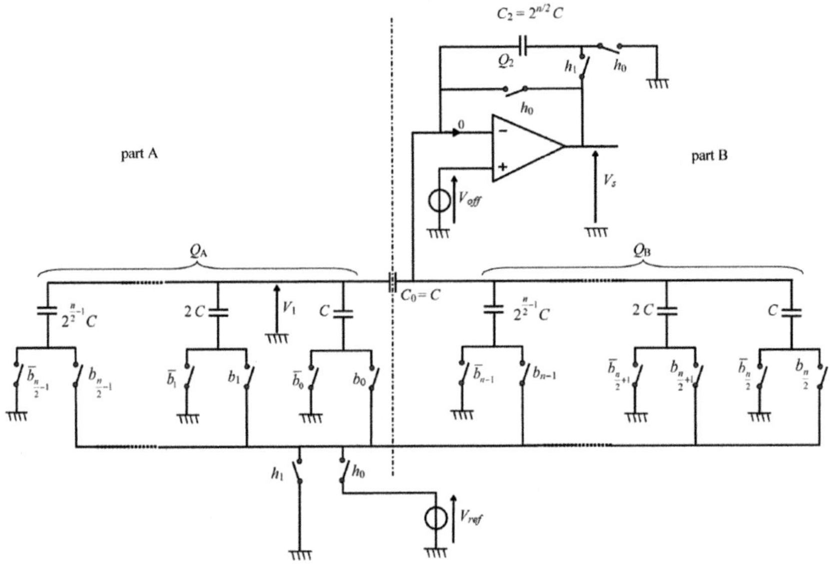

Figure 2.13. *Parallel converter based on charge redistribution and reduced capacitance*

The two sections are coupled by a capacitance C. The operation is based on the same sequence as previously seen. In the first phase, $h_0 = 1$ and $h_1 = 0$, inducing the memorization of the offset voltage of the operational amplifier into C_2 and the charge of every capacitance of the lower portion of the diagram by means of voltage $V_1 - b_k V_{ref}$ in part A and $V_{off} - b_k V_{ref}$ in part B. In the next phase, where $h_0 = 0$ and $h_1 = 1$, the conservation of charge must be considered separately in both parts A and B because the upper armatures of the capacitors of part A and the left armature of C_0 constitute a first isolated system, whereas the upper armatures of the capacitors of part B, the right armature of C_0 and the left armature of C_2 simultaneously constitute a second isolated system. By combining the results of the two redistributions of charge (see exercise 2.6.3), the sum of the contributions due to half of the bits of strongest weight and due to the other half of the bits of lowest weight is recovered in capacitor C_2. We thus arrive at the same result as before, with capacitances comprised within a greatly reduced range and thereby that provide a significant gain of speed. Under these conditions, it is possible to reach $n = 12$ bits, or even more, because the relationship between extreme capacitances is equal to $2^{n/2}$, which is here 64 and which is a much more reasonable value than $2^n = 4{,}096$.

2.3.4.3. *D-A mixed converter*

The conversion of groups of bits of different weight is, in this case, sequentially achieved by combining a parallel system that processes each group of p bits ($p < n$) and the addition of each successive result with a serial system in order to obtain the n-bit conversion (Figure 2.14). The intermediate storage is carried out in the capacitance $C_2 = C$, which no longer performs the division by 2^p, which is instead achieved through re-looping the ladder network of the p capacitors onto the output of the operational amplifier. The conversion starts with the p bits of lowest weights.

The initial sequence, $h_1 = 1$, $h_2 = 0$, $h_3 = 1$ (switches open if 0, closed if 1), yields $Q_{10} = -(2^{p-1}b_{p-1} + \ldots + b_0)CV_{ref} + (2^p-1)CV_{off}$ while the operational amplifier operates as a follower circuit and the capacitance C_2 stores the offset voltage V_{off}: $Q_{20} = CV_{off}$.

The second stage is composed of three phases:

(1) $h_1 = 0$, $h_2 = 1$, $h_3 = 0$, $b_k = 1\ \forall k$; allowing all of the ladder capacitors to be re-fed back onto the output of the operational amplifier since the switches b_k are all closed, which leads to $Q_{11} = -(2^p - 1)C(V_{s1} - V_{off})$ and $Q_{21} = -C(V_{s1} - V_{off})$. From the conservation of charge in the isolated part $Q_{10} + Q_{20} = Q_{11} + Q_{21}$, it follows that $V_{s1} = (b_{p-1} + \ldots + b_0/2^{p-1})(V_{ref}/2)$, which is stored in the capacitance C_2 because in the following phase, h_2 switch opens before h_1 switch closes.

(2) $h_1 = 1$, $h_2 = 0$, $h_3 = 0$ and the following p bits are transmitted over the ladder capacitance network, causing $Q_{12} = -(2^{p-1}b_{2p-1} + \ldots + b_p)CV_{ref} + (2^p-1)CV_{off}$ and $Q_{22} = -C(V_{s1} - V_{off})$ because capacitance C_2 being no longer connected, it preserves its charge.

(3) $h_1 = 0$, $h_2 = 1$, $h_3 = 0$, $b_k = 1\ \forall k$; configuration identical to that of the first phase. It results thereof that $Q_{13} = -(2^p - 1)C(V_{s3} - V_{off})$ and $Q_{23} = -C(V_{s3} - V_{off})$. The conservation of charge of the isolated part yields $Q_{12} + Q_{22} = Q_{13} + Q_{23}$, resulting in:

$$V_{s3} = \left(b_{2p-1} + \frac{b_{2p-2}}{2} + \cdots + \frac{b_p}{2^{p-1}} + \frac{b_{p-1}}{2^p} + \cdots + \frac{b_0}{2^{2p-1}} \right) \frac{V_{ref}}{2} = \frac{V_{ref}}{2^{2p}} \sum_{k=0}^{2p-1} 2^k b_k$$

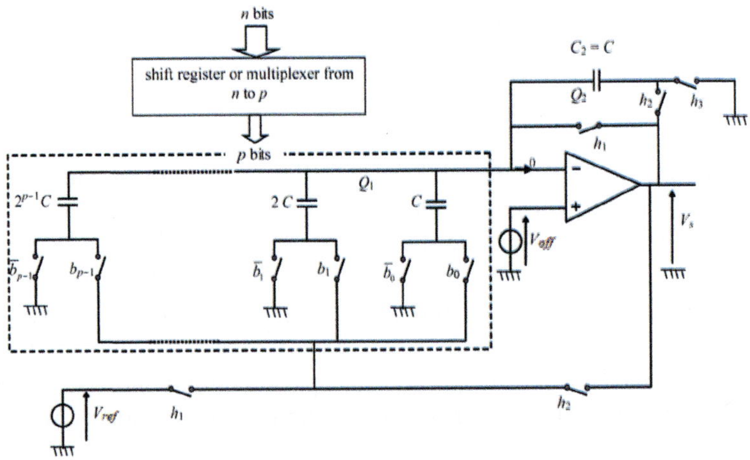

Figure 2.14. *Parallel-series DAC*

If $n = 2p$, the process is complete and the converted voltage is $V_{s3} = \frac{V_{ref}}{2^n}\sum_{k=0}^{n-1} 2^k b_k$ as in the previous DACs. Otherwise, the process continues restarting with phase 2. This conversion system is a good compromise between a reduced number of capacitances compared to the parallel DAC and a small number of steps compared to the series DAC.

2.4. Analog-to-digital conversion

The conversion between the analog quantity and a quantized quantity equal to M times the quantum q_0 is always performed by means of one or more comparators or an integrator. Therefore, it is desirable that the offset of the comparators or integrator be minimized, or better to compensate for it, in order to ensure the best conversion precision. This criterion is satisfied if the error resulting from the residual offset does not exceed ±1/2 LSB. The binary encoding is achieved with n bits.

2.4.1. *Converter using 2^n comparators or flash converter*

The comparison of the analog voltage V_a is performed with all the multiples of the quantum of voltage q_0, from 0 up to $(2^n - 1)\,q_0$ or rather from $q_0/2$ to $(2^n - 1/2)\, q_0$. This operation requires 2^n comparators (Figure 2.15).

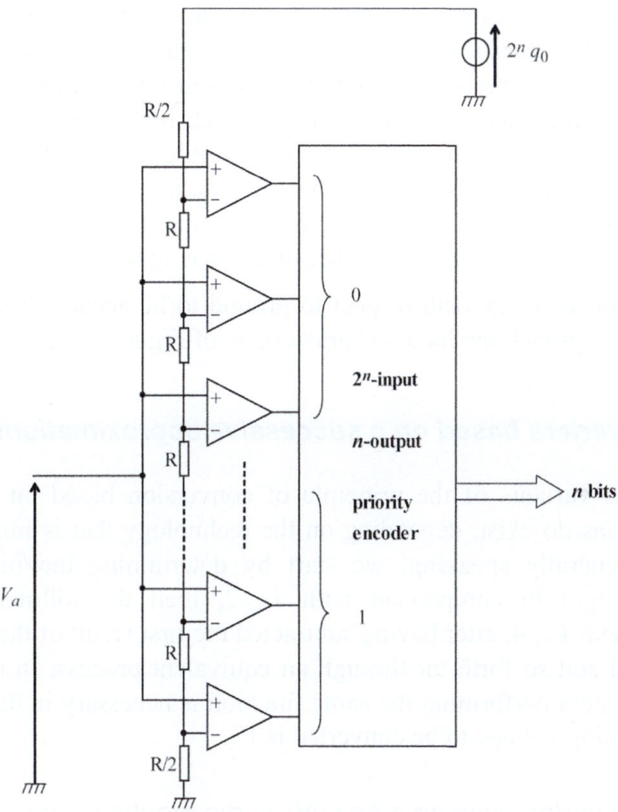

Figure 2.15. *Flash converter*

If $\left(N-\dfrac{1}{2}\right)q_0 < V_a < \left(N+\dfrac{1}{2}\right)q_0$, the first N outputs counted from the bottom are at logic level "1" (or H). A priority encoder (see Chapter 3 of Volume 1 [MUR 17a]), [NDJ 16a] then generates the number M encoded into n bits. The conversion time is limited only by the propagation times through the comparators and the encoder. At the onset of this technique, the number of bits did not exceed 6 or 7, but advances in integration and the improvement of comparators have made DACs of this type with 8 bits, or even 9 or 10 bits, much more widespread. On the one hand, there is no compensation of offset voltages, which must therefore be reduced to the lowest possible values. Moreover, on the other hand, 2^n input currents flow through the uppermost resistance, $2^n - 1$ flow through the second and so on, in addition to the current that goes to ground. Consequently, the identity of all the values of the resistances is not a solution that guarantees the same potential difference across the terminals of each resistor if the sum of the input currents of the comparators is non-negligible compared to the current flowing to ground. This issue had been overcome in a decisive manner when the technology for the adjustment of resistances by laser beam was implemented in integrated circuits, because it has made it possible for each of the 2^n node voltages with respect to ground to be accurately adjusted in order to best approach the ladder characteristic of Figure 2.1.

2.4.2. *Converters based on n successive approximations*

Numerous variants of the principle of conversion based on successive approximations do exist, depending on the technology that is implemented. However, generally speaking, we start by determining the bit with the strongest weight in comparison with $V_{ref}/2$, then the following bit by comparing with $V_{ref}/4$, after having subtracted the first result of the voltage to be converted and so forth, or through an equivalent process. In all cases, a DAC or a system performing the same function is necessary in the feedback loop. The analog voltage to be converted is V_a.

2.4.2.1. *Converter using an n-bit DAC in the feedback loop*

We use the natural binary code for $M = b_{n-1}\,2^{n-1} + b_{n-2}\,2^{n-2} + \ldots + b_k 2^k + \ldots + b_0$.

The first comparison allows us to determine (Figure 2.16):

b_{n-1} by comparing V_a to $2^{n-1}\,q_0$;

then b_{n-2} by comparing V_a to $(b_{n-1}\,2^{n-1} + 2^{n-2})\,q_0$;

then b_{n-3} by comparing V_a to $(b_{n-1}\,2^{n-1} + b_{n-2}\,2^{n-2} + 2^{n-3})\,q_0$;

and so on.

The n RSh flip-flops that memorize the b_k are controlled by the output of the comparator. The outputs of the sequencer made of a shift register are all 0 during the first conversion step except the first which is 1 (or H). This high level then shifts for each clock pulse Q_{n-1}, Q_{n-2} , Q_{n-3} and so on, successively requiring that $b_{n-1}, b_{n-2}, b_{n-3}$, etc., equal to 1. Then, the result C of the comparison enables $b_{n-1}, b_{n-2}, b_{n-3}, \ldots = 1$ if $V_a > V'_a$ (C = 0 does not modify the output of the flip-flop which remains at 1) or subsequently reestablishes $b_{n-1}, b_{n-2}, b_{n-3}, \ldots = 0$ if $V_a < V'_a$ (C = 1 resets the output of the flip-flop to 0).

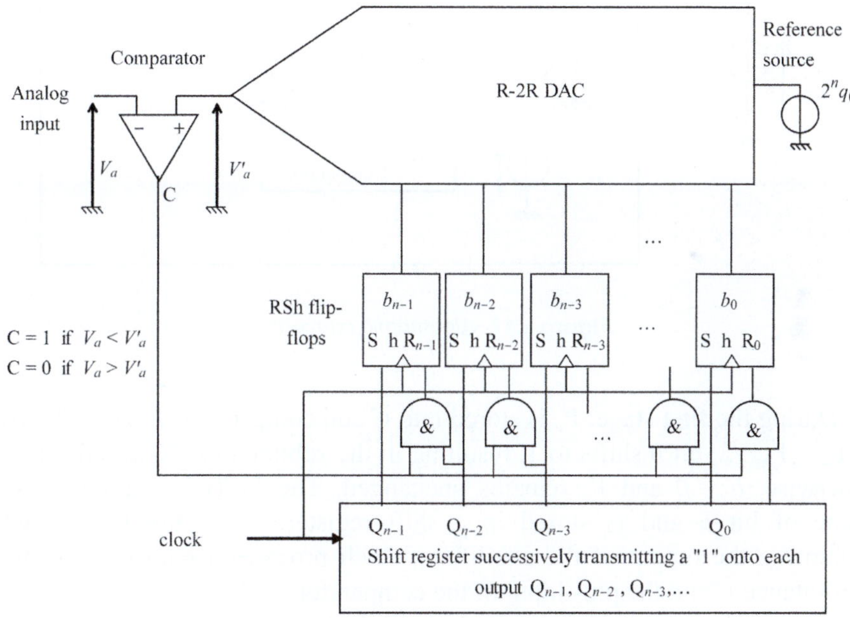

Figure 2.16. *Successive approximations converter*

The voltage V_a to be converted should be maintained constant for the duration of the conversion, hence the need for a sample-and-hold circuit on input.

This converter represents a good trade-off between the number of comparators (only 1) and the duration of conversion that lasts n clock cycles in addition to the zeroing (Reset). The R-2R DAC type is commonly used in the feedback loop but any other type can replace it as long as it computes the result of every comparison in due time.

2.4.2.2. *Algorithmic converter*

In this converter, at each step, the residual originating from the difference between the input voltage or the previous residual and $V_{ref}/2 = 2^{n-1}\ q_0$ is multiplied by 2 in order that it always be compared to $V_{ref}/2$ (Figure 2.17). This is still a sampled system based on sequential operation.

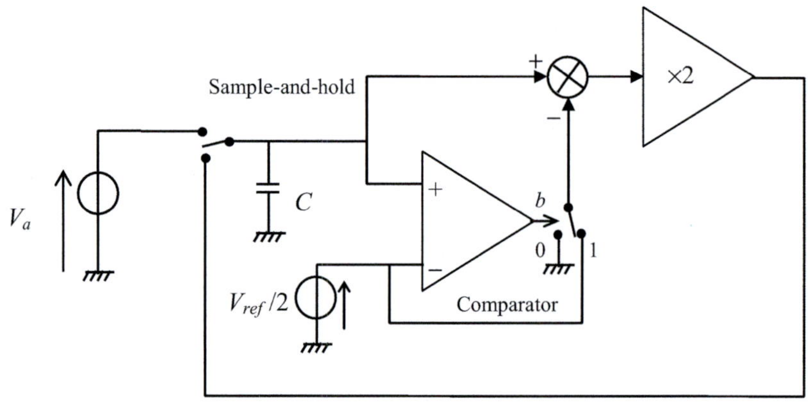

Figure 2.17. *Algorithmic converter*

During the first stage, V_a is stored into C and compared to $V_{ref}/2 = 2^{n-1}q_0$. If $V_a > V_{ref}/2$, bit b shifts to 1, resulting in the subtraction of $V_{ref}/2$ from V_a; otherwise, $b = 0$ and V_a remains unchanged. The MSB b_{n-1} assumes the value of bit b and is stored in a shift register. Then, the first switch commutes the output of the amplifier (which provides a gain of 2) to the capacitance C and the plus input of the comparator.

During the second stage, the residual $V_a - b_{n-1} V_{ref}/2$ is multiplied by 2 and again compared to $V_{ref}/2$. This is tantamount to comparing the residual $V_a - b_{n-1} V_{ref}/2$ to $V_{ref}/4$. The result will yield b_{n-2}, which will be used to compute $2\times(V_a - b_{n-1} V_{ref}/2 - b_{n-2} V_{ref}/4)$. This new residual will be multiplied by 2 and the cycle will continue until the LSB is determined. The voltage offset of the comparator can be generally neglected because the signal level on its plus input is always of a magnitude comparable to $V_{ref}/2$, due to the multiplication by 2 of the residual at each step.

Each bit is successively placed into an n-stage shift register and all the bits are available after n steps. In the event that it is desirable to achieve a result of the conversion for every step, this is possible using a "pipeline" structure which consists of transferring the residual into an equivalent system instead of feeding it back. There are then n stages in cascade, each delivering bits b_{n-1}, b_{n-2} , b_{n-3} and so on, to the output of each comparator. The result of the conversion is updated during each step provided that after each of these outputs a shift register is added with $n-1$ cells for b_{n-1}, $n-2$ cells for b_{n-2}, $n-3$ cells for b_{n-3} and so forth. Therefore, once the first conversion is completely exhausted, the delay caused by each shift register still allows for all the bits related to a conversion to reach at the same time the last cell of each register. Although the conversion time is always that of n steps, the updating rate is the same as the time devoted to a single stage, which is tantamount to maintaining the same sampling rate.

2.4.3. *Mixed or semi-flash converter*

This converter employs the principles of previous ones but only utilizes two stages in the process because the feedback loop comprises an n/2-bit DAC (Figure 2.18).

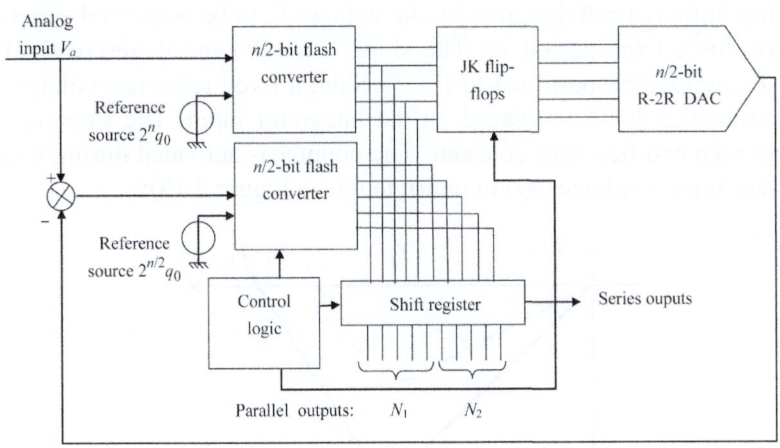

Figure 2.18. *Mixed converter*

The first stage enables the n/2 bits with strongest weights to be obtained as well as outputs N_1 such that the analog input voltage V_a is approximated by $N_1 q_0$ within an uncertainty of $2q_0$. The second stage implements the conversion of the $n/2$ bits of lowest weights, starting from the analog residual computed by subtraction of $N_1 q_0$ from V_a and outputs $N_2\, q_0$ within an uncertainty of q_0. The full conversion result $M = N_1 + N_2$ is built by the means of simple shifting subsequent bits in a register.

This system proves very advantageous since, for an n-bit conversion, it requires only two times $2^{n/2}$ comparators: n = 10 bits would give 1,024 comparators for a normal flash converter, whereas only 64 are needed for the mixed converter; moreover, it remains very fast since only two cycles and zero resetting are needed.

2.4.4. *Ramp converters*

Ramp converters all make use of an integrator of time constant RC and a comparator that validates a counter for a time proportional to the voltage to be converted. In the simple ramp category, the analog voltage is directly compared with a ramp. Nonetheless, this system exhibits too many inaccuracies and has been replaced by the double ramp described hereafter. The voltage $s(t)$ is the one existing at the output of an integrator which the input has initially been driven to by the voltage V_a to be converted, supposed positive, for a fixed period T_1. The slope $ds(t)/dt$ and therefore the final value V_1 are then proportional to V_a. Second, a fixed reference voltage V_{ref}, of opposite sign to V_a, is placed on the integrator input. The slope $ds(t)/dt$ changes sign and becomes constant. The counter is activated during time T_2 necessary for the voltage $s(t)$ to return to 0 (see Figure 2.19):

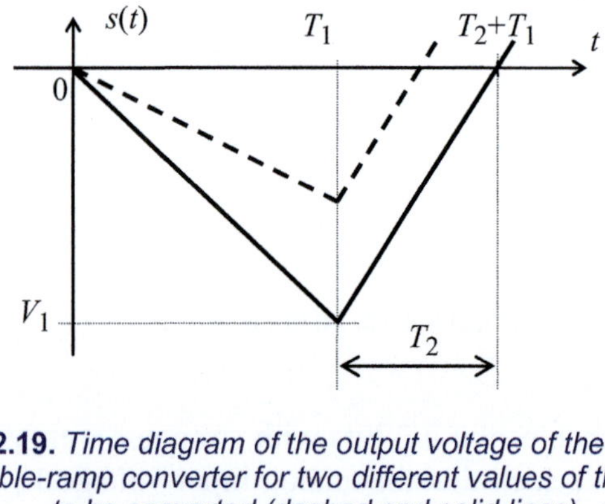

Figure 2.19. *Time diagram of the output voltage of the integrator in a double-ramp converter for two different values of the voltage to be converted (dashed and solid lines)*

1) After the fixed time T_1 : $s(T_1) = V_1 = -\frac{V_a}{RC}T_1$.

2) Next, the input of the integrator is switched onto V_{ref}(here $V_{ref} < 0$) and the zero crossing of $s(t)$ determines the time $T_2 \Rightarrow V_1 = \frac{V_{ref}}{RC}T_2$.

By eliminating V_1, the count time is T_2 from which it follows that $V_a = \left|V_{ref}\right|\frac{T_2}{T_1}$.

This method is very accurate (at least up to 22 bits with integrated circuit) because the counted time depends only on the clock frequency and the reference voltage, once the errors introduced by the operational amplifier have been corrected. It is however very slow because it requires a fixed time period T_1 plus 2^n clock ticks maximum. Conversion time is then comprised between a few milliseconds and a few hundred milliseconds, which is acceptable for digital multimeters only.

When high precision is required (12 bits and more), it is necessary to correct the errors. The most important originates from the offset error of the integrator amplifier. A previous sequence is then added during which its input is switched to ground and the output is measured, which is non-zero due to the offsets. This preliminary measure is stored and then it is subtracted from the value measured by the double-ramp method in one or two steps, respectively, leading to triple and quadruple ramp converters.

2.5. "Sigma-delta" conversions

"Sigma-delta" conversions implement analog and digital operations that borrow most of their modeling from the field of sampled systems (see Chapter 1). However, the entire process is rather sophisticated in order to satisfy a requirement that consists in finding a compromise between the reduction of the number and of the complexity of functions being employed, as well as a minimal degradation of the signal-to-noise ratio which is obtained by shifting the quantization noise spectrum toward frequencies where it can be more easily filtered out. This process is based on a servo-control system called a modulator, whose function is to reduce the error between the input signal and the average value of the converted signal. This modulator includes an integrator providing a sum of samples, hence the term "sigma", then a difference to calculate an error signal, hence the term "delta."

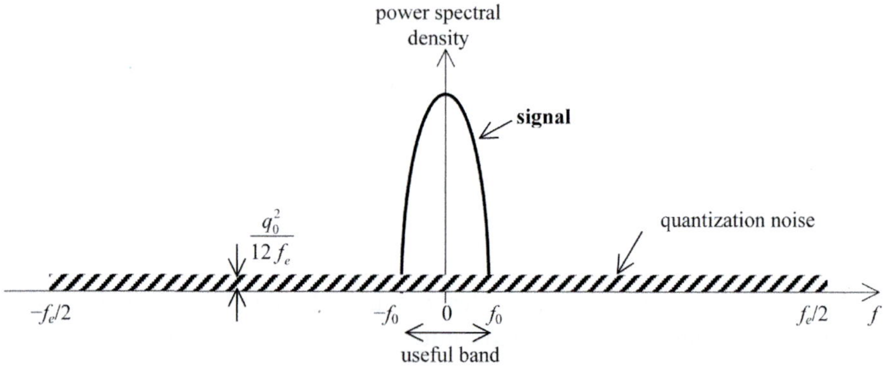

Figure 2.20. *Power spectra of the signal and of the noise sampled in a sigma-delta modulator (the satellites of the central spectrum of the signal are not visible because they are located around integer multiples of the sampling frequency f_e, which can be found outside of the graph)*

The process initially consists of degrading the conversion resolution, which increases the quantization noise, and at the same time in increasing the sampling rate of the signal in order to oversample it in the same proportions as the ratio between the quantum used in the modulator and the one targeted on process completion (Figure 2.20). This ratio is also that appearing between the oversampling frequency $f_e = 1/T_e$ and the Nyquist frequency just above f_0 that is necessary for the sampling theorem to be verified at the analog signal level.

The presentation will begin with that of an ADC based on elementary functions allowing for the development of a first-order modulator and will then proceed by generalizing the study by means of models of sampled systems capable of addressing higher-order modulators and of distinguishing between signal processing and noise processing.

2.5.1. *Basic first-order modulator-based "sigma-delta" ADC*

A "sigma-delta" ADC outputting a result encoded into n bits (Figure 2.21) can be designed based on the idea that the ramp converter (section 2.4.4) can be improved by replacing the count from 0 by a readjustment of the result according to the difference between the old and the new converted or counted value. This thus avoids the need to restart counting from 0 at each conversion by implementing a control able to follow the voltage V_a to be converted up to the quantization error. This causes a much faster updating of the comparison voltage (here V_s in Figure 2.21) and of the result of the counting. This is so because it only requires a number of clock ticks limited to generating a ramp between the old and the new value of the voltage to be converted when the latter has changed. When the voltage to be converted remains constant, the count fluctuates regularly within a single unity and the integrated voltage V_i fluctuates around V_a. The system implemented with a single integrator is composed of a first order sigma-delta modulator processing m bits ($m << n$) and of a decimation filter (or decimator) placed outside and following the modulator.

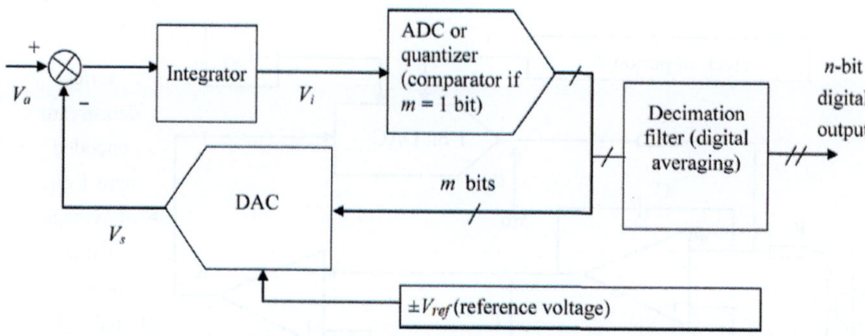

Figure 2.21. *Block diagram of a first order sigma-delta converter*

The decimation filter is both an averaging and a low-pass filter that restores the n-bit resolution by eliminating the quantization noise whose spectrum has been translated toward higher frequencies by means of oversampling, and simultaneously an under-sampler, because it reduces the data output rate with respect to the clock frequency f_e of the sigma-delta modulator. The low-pass function ensures the validity of the sampling theorem (see Chapter 1) and the filtering of the quantization noise. It simultaneously allows us to reestablish an encoding onto a number of bits $n >> m$, and for all these reasons, it necessarily processes data in digital form. Its properties, already examined in the previous chapter, will be specified further in the case of this application. The arrangement can be described as shown in Figure 2.21. The integrator in the loop is intended to cancel the static error. The number of bits m is much lower than that eventually obtained (n) for the conversion, but this is offset by an increase in the rate of conversions (oversampling).

In practice, we often had <u>$m = 1$ bit</u> when this technology was originally developed for simplification reasons. This case is addressed in the first place. The ADC is then a simple comparator, which is also a quantifier, and that detects the sign of the voltage error V_{err}. Furthermore, the DAC provides a voltage equal to $\pm V_{ref}$ depending on this sign (Figure 2.22). After N clock ticks (more specifically a time interval NT_e if T_e is the sampling period) $NV_a + (M_1 - M_2)V_{ref}$ is approximately 0 due to the control loop, with M_1 = number of periods where $+V_{ref}$ has been applied and M_2 = number of periods where $-V_{ref}$ has been applied (and $N = M_1 + M_2$).

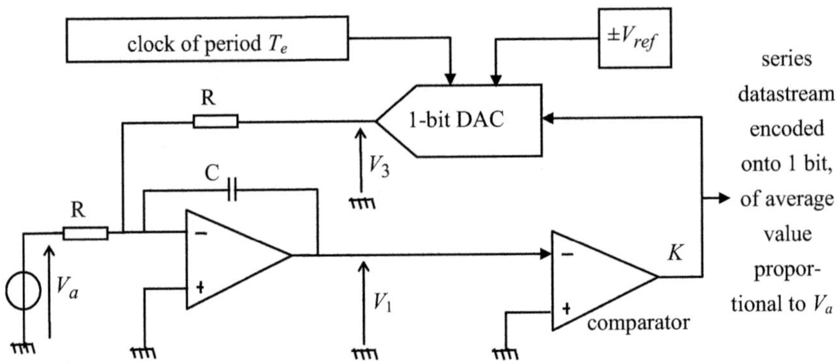

Figure 2.22. *Sigma-delta converter with 1-bit DAC*

We can deduce thereof that $V_a = \frac{M_2 - M_1}{N} V_{ref}$ (see exercise 2.6.4), which makes it possible to obtain the result of the conversion $M_2 - M_1 = N \frac{V_a}{V_{ref}}$ on the output of an up–down counter activated by the binary variable K whose value determines the value output by the DAC. This difference is nothing else than the algebraic value of the integral of the difference between the analog voltage and that generated by the feedback loop of the modulator over a time interval of N clock periods T_e. The M_2 low states and M_1 high states being alternated in a random fashion on the output of the comparator, the average value on the output of the DAC is equal to V_a but a conversion noise is superimposed thereto whose spectrum is spread as indicated in Figure 2.20. The flow of binary data (*bitstream*) on the output of the comparator is encoded onto 1 bit and must be converted into n-bit-encoded words by the decimation filter which also plays the role of attenuating the quantization noise. This system being based on the 1-bit modulator is in addition particularly simple and accurate if the accuracy and stability of reference sources is properly considered. Conversion times of a few microseconds are attained as well as effective resolutions from n =12 to 24 bits after the decimation filter.

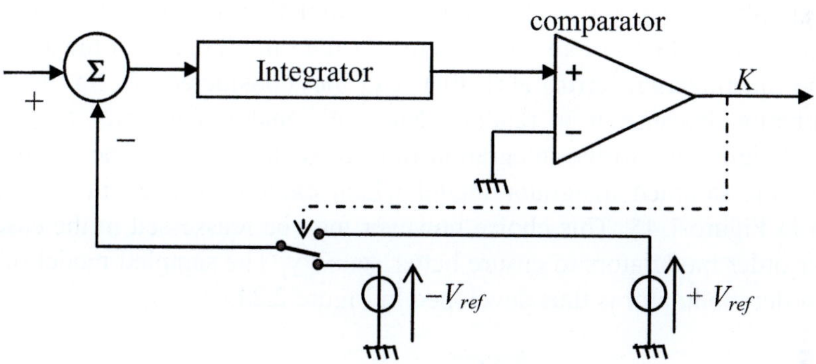

Figure 2.23. *1-bit modulator*

If only the part comprising a negative feedback loop, known as "1-bit modulator", is taken into consideration it may be more concisely represented in Figure 2.23. When it is desirable to obtain an analog voltage, one can directly utilize the output of the 1-bit modulator or the signal obtained after the switch in the feedback loop (Figure 2.23) insofar as the average value of states 1 and 0 will be representative of the input voltage if one considers the average error produced by the subtractor to be negligible, as already previously assumed. At the minus input of the subtractor, the result is a rectangular pulse voltage $\pm V_{ref}$ whose time periods at states $+ V_{ref}$, or $-V_{ref}$ are such that on average we get a voltage $\frac{M_2 - M_1}{N} V_{ref}$ equal to V_a. This can be regarded as some kind of pulse width modulation (PWM). This operation justifies the use of the name "modulator" for these subsets, a designation that is kept for multi-bit systems with m >1.

2.5.2. *First-order modulator sampled model*

Insofar as modulators equally include analog and logical or digital functions, the description of these systems in the discrete-time domain and using the z-transform (ZT) is absolutely appropriate since the analog or digital nature of variables and functions are not really relevant in this representation. One of the integrators already represented in Figure 1.45, can be found as distinct functions in the loop of the first-order modulator of Figure 2.21, the ADC and the DAC. The function of these last two elements is to simply convert the analog signal into digital signal or vice versa, and consequently it does not alter by any means the samples, except the addition of the quantization error $e[k]$ that can be considered as an external perturbation because it is random. Since an analog integrator outputs a delayed signal due to the integration time constant, it makes more sense to choose the sampled integrator model which exhibits a delay in the direct chain in Figure 1.45. This choice however may be reassessed in the case of higher order modulators to ensure better stability. The sampled model of the first-order modulator is thus developed in Figure 2.24.

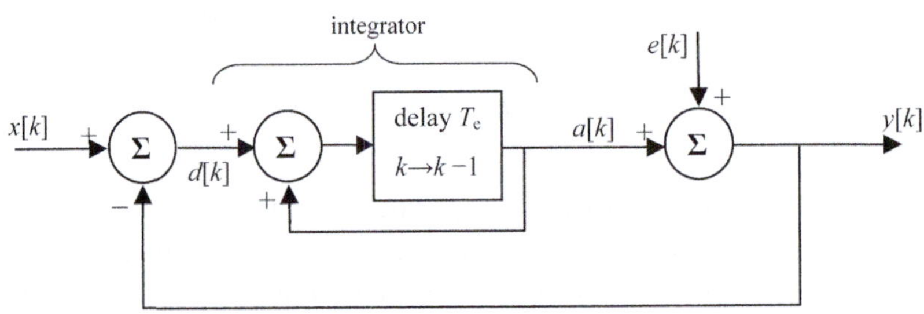

Figure 2.24. *First-order modulator sampled model*

The input sample is denoted by x, the output sample by y, the error due to the quantization noise by e, by d the difference $x-y$ and by a the sample on the integrator output. The sampling frequency is f_e ($=1/T_e$) and k the time index. Like in Figure 1.45. $a[k] = a[k-1] + d[k-1]$ for the integrator.

By traveling the loop, $y[k] = a[k] + e[k]$, and thus also $y[k-1] = a[k-1] + e[k-1]$, then $d[k] = x[k] - y[k]$, thus also $d[k-1] = x[k-1] - y[k-1]$.

From which is deduced by eliminating $a[k-1]$ and $d[k-1]$:

$y[k] = a[k-1] + d[k-1] + e[k] = y[k-1] - e[k-1] + d[k-1] + e[k] = y[k-1] - e[k-1] + x[k-1] - y[k-1] + e[k]$. That is finally:

$$y[k] = x[k-1] + e[k] - e[k-1],$$

whose ZT is $Y(z) = z^{-1}X(z) + (1 - z^{-1})\,E(z)$.

The output sample of the closed loop is therefore that of the input delayed by a period T_e added to the differentiated quantization error, the respective transfer functions being $H_{S1}(z) = z^{-1}$ for the signal and $H_{N1}(z) = (1 - z^{-1})$ for the noise.

The effect of this differentiation on the effective noise spectral density $\dfrac{q_0}{2\sqrt{3f_e}}$ in the frequency domain is that of a multiplication by a factor $\left|1 - \exp\left(-j2\pi\dfrac{f}{f_e}\right)\right|$, which increases from 0 up to 2 in the frequency interval $[0, f_e/2]$. This property is interesting because the noise spectrum of Figure 2.20 is modified on two different levels, favorable to improve the signal-to-noise ratio: in the useful band, the noise is attenuated because $\left|1 - \exp\left(-j2\pi\dfrac{f}{f_e}\right)\right|$ is largely smaller than 1, whereas it increases in the high-frequency domain when approaching $f_e/2$ where it can be effectively filtered. The noise power spectral density becomes $N_1(f) = \left|1 - \exp\left(-j2\pi\dfrac{f}{f_e}\right)\right|^2 \dfrac{q_0^2}{12f_e} = \dfrac{q_0^2}{3f_e}\sin^2\left(\pi\dfrac{f}{f_e}\right)$ which is zero for $f = 0$, then increasing and reaching its maximum for $f = f_e/2$.

2.5.3. *Modulators of order l > 1 and signal-to-noise ratio*

In order to build a sampled second-order modulator, an integrator is combined without delay with an integrator equipped with delay and the quantifier (Figure 2.25).

The transfer function of the first is $\dfrac{1}{1-z^{-1}}$ and the second $\dfrac{z^{-1}}{1-z^{-1}}$ (see Figure 1.52). We can then process the ZTs of the sampled signals. The outputs of the first and the second integrators are, respectively, designated by $A_1(z)$ and $A_2(z)$ while the inputs and outputs of the modulator are, respectively, called $X(z)$ and $Y(z)$. $E(z)$ is the quantization error generated inside the modulator loop in the z variable domain.

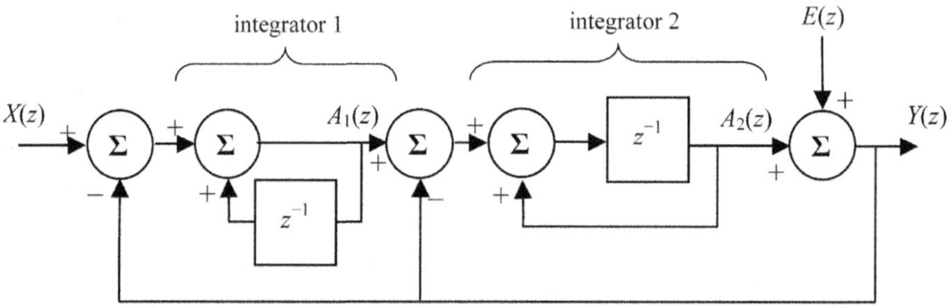

Figure 2.25. *Second-order modulator*

The transfer function for the signal follows from $Y(z)=A_2(z)=\dfrac{z^{-1}}{1-z^{-1}}\left(A_1(z)-Y(z)\right)$, where $A_1(z)=\dfrac{1}{1-z^{-1}}\left(X(z)-Y(z)\right)$.

By eliminating $A_1(z)$, it follows that $H_{S2}(z)=\dfrac{Y(z)}{X(z)}=z^{-1}$ that indicates that the transmittance is not modified for the signal compared to the first-order modulator. The output signal is simply delayed with respect to the input signal.

For the noise, $Y_{N2}(z)=A_2(z)+E(z)$. Since there is no noise on input, and we still have $A_2(z)=\dfrac{z^{-1}}{1-z^{-1}}\left(A_1(z)-Y_{N2}(z)\right)$, we finally get $Y_{N2}(z)=\left(1-z^{-1}\right)^2 E(z)$.

Therefore, the noise power spectral density is differentiated twice and becomes $N_2(f)=\dfrac{q_0^2}{12f_e}\left[2\sin\left(\pi\dfrac{f}{f_e}\right)\right]^4$.

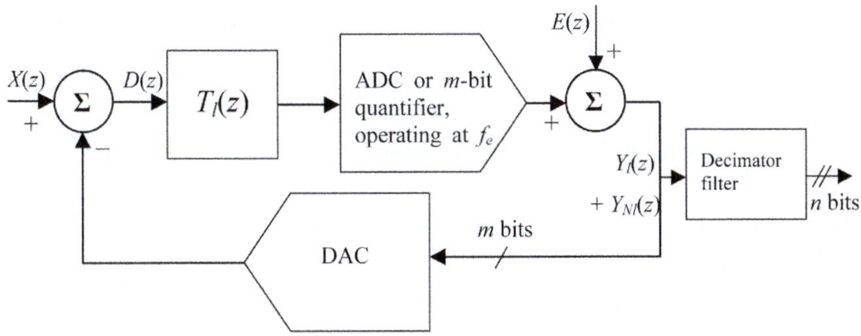

Figure 2.26. *Sampled model of a modulator of order l followed by a decimation filter*

Generally speaking, if a sampled transfer function $T_l(z)$ acts upon the signal $D(z)$ resulting from the difference between the input signal $X(z)$ and that of the feedback loop (Figure 2.26), the block diagram comprises in addition to $T_l(z)$ only the additive perturbation due to the quantization error $E(z)$. This is due to the fact that the DAC and the ADC only change the nature of the signals, analog or digital, but not their sampled representation, and thus present a transmittance equal to 1. Accordingly, the relations applicable to Figure 2.26 are for the signal $Y_l(z) = T_l(z)\,(X(z) - Y_l(z))$, which gives $H_{Sl}(z) = \dfrac{Y_l(z)}{X(z)} = \dfrac{T_l(z)}{1+T_l(z)}$, and for the noise $Y_{Nl}(z) = -\,T_l(z)\,Y_{Nl}(z) + E(z)$, which results in $Y_{Nl}(z) = \dfrac{E(z)}{1+T_l(z)}$.

If $\dfrac{1}{1+T_l(z)} = \left(1-z^{-1}\right)^l$, then $T_l(z) = \dfrac{1-\left(1-z^{-1}\right)^l}{\left(1-z^{-1}\right)^l}$ and the closed-loop gain becomes $\dfrac{T_l(z)}{1+T_l(z)} = 1-\left(1-z^{-1}\right)^l = z^{-1}\sum_{i=1}^{l}(-1)^{k-1}C_l^k\,z^{-(k-1)}$,which indicates that the output signal is composed of a linear combination of samples delayed for one or more sampling periods while the noise is differentiated l times. Nonetheless, this applies for the signal only in the case of a single feedback loop, which is not systematically implemented, and the example of the previous second-order modulator shows that it is possible to obtain a simple delay of a single sampling period for the signal while differentiating twice the noise.

We can thus generalize the output noise of a modulator of order l by $Y_{Nl}(z)=\left(1-z^{-1}\right)^{l}E(z)$. The noise power spectral density then becomes $N_l(f)=\frac{q_0^2}{12f_e}\left[2\sin\left(\pi\frac{f}{f_e}\right)\right]^{2l}$. The transposition of the noise power located in the low frequency range, inside the useful band, toward high frequencies close to $f_e/2$ is therefore all the more effective if the order of the modulator is higher because the sine function is close to $\pi f/f_e$ when $\pi f/f_e$ <<1, it is all smaller when l is larger in the useful band where $f<f_0$ as can be observed in Figure 2.27. It is therefore possible to significantly improve the signal-to-noise ratio in the useful band, the excess noise in the high frequencies being filtered after the modulator. In the approximation of a sampling rate f_e significantly higher than the frequency f_0 which limits the useful band, the noise power in the useful band can be calculated by replacing the sine by its argument:

$$P_{Nl}\approx\frac{(2\pi)^{2l}q_0^2}{12f_e^{2l+1}}\int_{-f_0}^{+f_0}f^{2l}\,df=\frac{\pi^{2l}q_0^2}{12(2l+1)}\left(\frac{2f_0}{f_e}\right)^{2l+1}\quad \text{for } f_0<<f_e/2.$$

The exact function (without the sine approximation) is plotted in Figure 2.27.

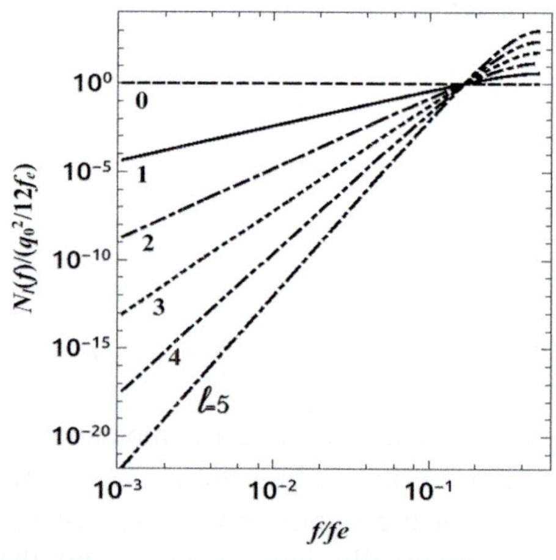

Figure 2.27. *Noise power spectral density referenced to the standard value according to the frequency normalized by the sampling frequency, for different values of the order l of the modulator (0th order corresponds to a quantization noise spectral density which is original and remains unaltered)*

We can then evaluate the signal-to-noise ratio when an m-bit quantifier is employed, offering the possibility of converting a signal of maximal amplitude over 2^m levels equidistant from q_0, which corresponds to an RMS value $S_{\text{rms}} = \dfrac{2^m q_0}{2\sqrt{2}}$ in the case of a sinusoidal signal.

Under these circumstances, the maximal signal-to-noise ratio in decibels is written by substituting P_{NI} by its expression previously demonstrated:

$$10\log\left(\frac{S_{\text{rms}}^2}{P_{NI}}\right) \approx (20l+10)\log\left(\frac{f_e}{2f_0}\right) + 20m\log 2 - 10\log\left(\frac{2\pi^{2l}}{3(2l+1)}\right).$$

It can be observed that the order of the modulator and the ratio $R = \dfrac{f_e}{2f_0}$ (of the Nyquist frequency to the frequency limiting the useful band) are crucial to improve the signal-to-noise ratio (Figure 2.28) as well as the number of bits of the quantizer to a lesser extent, because of the second term that corresponds to an increase of $6m$ dB. This is the reason why it is preferable to adopt a value m = 1 bit that requires a simple comparator operating as a quantifier (Figure 2.23) every time the technology makes it possible to reach a value of f_e sufficient to maximize the first term, the third one having a much smaller impact.

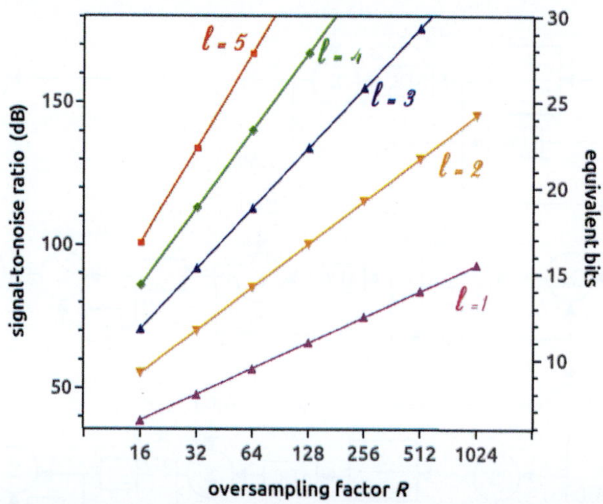

Figure 2.28. *Maximal signal-to-noise ratio for a lth-order 1-bit modulator; for an m-bit modulator, the curves would be shifted up by 20 log m in dB or m equivalent bits. For a color version of this figure, see www.iste.co.uk/muret/electronics3.zip*

As a matter of fact, it can be observed in Figure 2.28 that, with $m = 1$ bit, we can increase the signal-to-noise ratio by doubling the ratio R, of a number of equivalent bits (i.e. also 6 dB per equivalent bit) all the more so when a higher-order modulator is used, while it is increased by only 1 equivalent bit (or 6 dB) by doubling the number of levels, that is equivalently by increasing m by 1 bit. This justifies our aim to utilize a modulator of order higher than two, especially when the purpose is to reach 24 equivalent bits, which is now a current value in numerous applications. However, since we are faced with stability problems if the block diagram used for the second-order modulator is extrapolated to a system with three feedback loops or more, it becomes important that other solutions be found, as examined in the next paragraph.

2.5.4. *Stable modulators of order greater than two and CMOS technology-based circuitry*

Multiple feedback can be avoided by separating the noise processing implemented by several stages of the first-order modulator in cascade, from signal processing implemented by a single first- or second-order modulator. This is what is achieved in the modulator using several noise shaping stages (MASH = "multi-stage noise shaping", Figure 2.29).

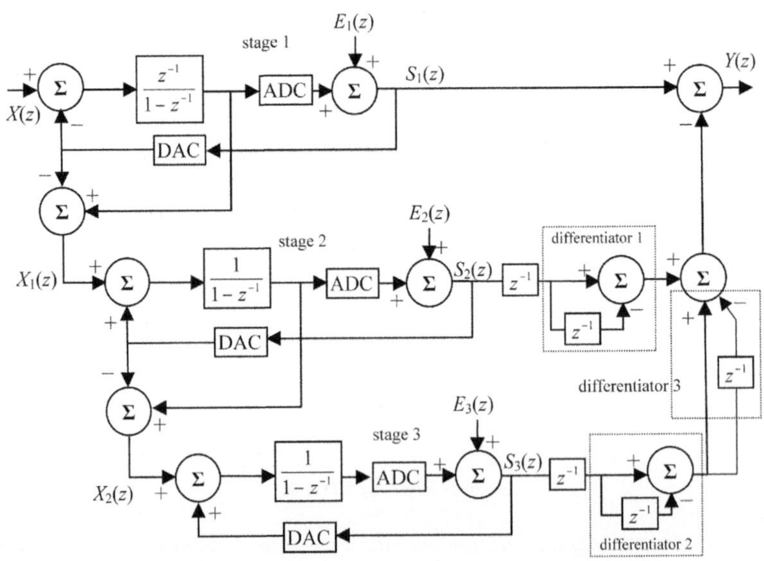

Figure 2.29. *Third-order "MASH" modulator*

At the end of every stage, with the exception of the last, the sum of the reconverted analog signal and of the quantization noise present in the loop is subtracted from the output signal of the integrator, which results in every following stage processing only the noise from the previous stage. The resulting noise is then delayed and differentiated $i-1$ times on the output of each stage numbered i and finally subtracted from the output signal of the first stage. Consequently, with the notations of Figure 2.29 and according to what has been already determined for the second-order modulator, it can be shown that (see exercise 2.6.5):

$$S_1(z) = z^{-1}X(z) + (1 - z^{-1})\,E_1(z) \; ; \quad X_1(z) = -E_1(z) \; ; \; X_2(z) = -E_2(z)$$

and finally $Y(z) = z^{-1}X(z) + (1 - z^{-1})^3\,E_3(z)$.

Therefore it follows that only the noise from the third stage differentiated three times is added to the digitized signal. It should be noted that all delays (of transmittance z^{-1}) illustrated in Figure 2.29 act upon digitized signals. It is thus necessary to implement them with a number of shift registers equal to the number of bits. In addition, adders and subtractors that follow must be able to process incoming and outgoing carryovers and to output signed binary words with the sufficient number of significant bits, which increases along with the summations or subtractions (see Chapter 3 of Volume 1 [MUR 17a] for details about these numbering techniques). The scattering of components may restrict the applicability of this technique to the fourth order at most.

Another system is based on multiple loops capable of developing a linear combination of analog samples of the signal after l integrator stages while the comparison with the converted signal is performed only once, which also generates quantification noise in a single loop only (Figure 2.30). The integrators introduce no delay, whereas conversion by the ADC is accompanied by a delay.

This system, named interpolative modulator, is flexible insofar as the adjustment of the coefficients of the signal and of the noise transfer functions provide a means to ensure stability and to shape the noise spectrum. According to Figure 2.30, the signals can be expressed as follows:

$$Y(z) = z^{-1}\left\{\sum_{i=0}^{l}\frac{a_i}{(1-z^{-1})^i}\big(X(z)+D(z)-Y(z)\big)\right\}$$

and $D(z)=\sum_{i=0}^{l}\frac{b_i}{(1-z^{-1})^i}\left(X(z)+D(z)-Y(z)\right)$

or also $D(z)=\dfrac{\sum_{i=1}^{l}\frac{b_i}{(1-z^{-1})^i}\left(X(z)-Y(z)\right)}{1-\sum_{i=1}^{l}\frac{b_i}{(1-z^{-1})^i}}$.

From which it can be derived that

$$Y(z)=\frac{z^{-1}\sum_{i=0}^{l}\frac{a_i}{(1-z^{-1})^i}}{1-\sum_{i=1}^{l}\frac{b_i}{(1-z^{-1})^i}+z^{-1}\sum_{i=0}^{l}\frac{a_i}{(1-z^{-1})^i}}X(z)=H_{itp,l}(z)X(z),$$

where the signal transfer function is $H_{itp,l}(z)$. For the noise $N_{itp,l}(z)$ on output:

$$N_{itp,l}(z)=z^{-1}\left\{E(z)+\sum_{i=0}^{l}\frac{a_i}{(1-z^{-1})^i}\left(D'(z)-N_{itp,l}(z)\right)\right\},$$

with $D'(z)=-N_{itp,l}(z)\dfrac{\sum_{i=1}^{l}\frac{b_i}{(1-z^{-1})^i}}{1-\sum_{i=1}^{l}\frac{b_i}{(1-z^{-1})^i}}$,

which finally yields:

$$N_{itp,l}(z)=\frac{z^{-1}\left[1-\sum_{i=1}^{l}\frac{b_i}{(1-z^{-1})^i}\right]}{1-\sum_{i=1}^{l}\frac{b_i}{(1-z^{-1})^i}+z^{-1}\sum_{i=0}^{l}\frac{a_i}{(1-z^{-1})^i}}E(z)=H_{N,l}(z)E(z)$$

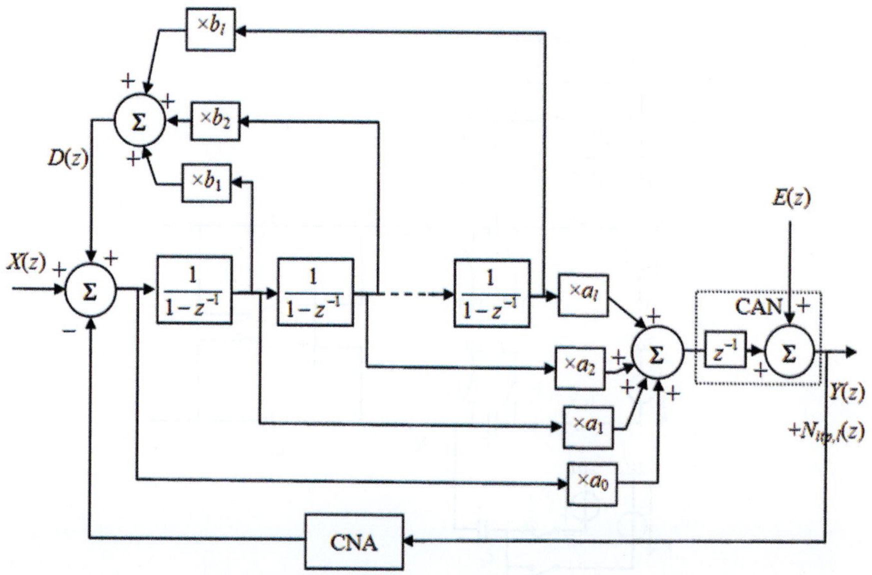

Figure 2.30. *lth-order-interpolative modulator*

The denominator being the same in both cases, the coefficients a_i and b_i allow us to respectively adjust the signal transfer function $H_{itp,l}(z)$ and that of the noise $H_{itp,l}(z)$, which verify $z\,H_{N,l}(z) + H_{itp,l}(z) = 1$. In order to benefit from an improvement of the signal-to-noise ratio comparable to that of the previous system, it is necessary that $H_{N,l}(z)$ be a lth-order high-pass transmittance and $H_{itp,l}(z)$ a lth-order low-pass transmittance. To achieve this goal, we can use a method for the synthesis of lth-order infinite impulse response filters and verify the result by simulation. Stability will be maintained if no pole of the transfer function is located outside the circle of radius 1 in the plane of the z variable. This system is better suited than the previous one when converters make use of a resolution higher than 1 bit because the number of converters is only two; the complexity however is much more significant in the analog part comprising numerous integrators, adders and amplifiers for adjusting the coefficients. This analog portion can be entirely designed using switched-capacitor circuits, and technology based on CMOS circuits, is thereby appropriate.

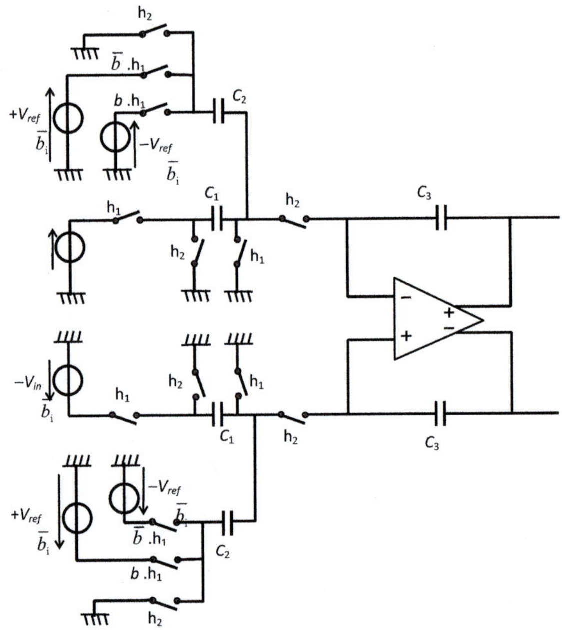

Figure 2.31. *First stage of a second-order switched-capacitor modulator (the variables h_1 and h_2 for the control of the switches are exchanged in the second stage)*

The "sigma-delta" technique is dependent on the implementation of modulators processing the sampled analog signal, most often achieved using switched-capacitor circuits. In Figure 2.31, a typical example of the first stage of a second-order modulator is schematized. The second stage is identical and cascaded after the first one but with the switches control variables h_1 and h_2 inverted. The assembly is symmetric, which makes it possible to eliminate the influence of the variations in reference voltage and injections of charges by the CMOS switches. The dependence of capacitances vis-a-vis the voltage is also eliminated at the first order. Two additional h_1 and h_2 clocks control the switches which themselves determine the charge transfers. Bit b is the one output by the comparator according to the sign of the voltage on its input, which follows the second stage. Each stage operates as an integrator, as can be observed by comparing with the exercises of Chapter 1. In total, it can be demonstrated that the sampled transfer function is indeed that of a second-order modulator such as used in section 2.5.3.

2.5.5. *Decimation filter*

The decimation filter provides several functions that can be understood by examining signal and noise spectra (Figures 2.20 and 2.27), the number of bits on the modulator output and by taking into account the decimation operation that consists of dividing the sampling frequency f_e utilized in the modulator by a number N much larger than 1, such that to reduce the output sampling frequency in the neighborhood of the Nyquist frequency, relative to the analog input signal. First, it must attenuate the portion of the spectrum situated outside the useful band, which, on the one hand, is due to the noise spectrum transposed to the frequencies close to $\pm f_e/2$ and its odd multiples, and, on the other hand, to the components of the output signal of the modulator outside the useful band that could possibly exist. In effect, the signal spectrum shown in Figure 2.20 is supposed to be already devoid of its components outside the useful band so that the sampling theorem be verified. Otherwise, this spectrum would be affected by the aliasing of these out-of-band components and thereby modified. This condition for the validity of the sampling theorem should be completed by prior filtering, but which is often considered as being superfluous if the out-of-band components are certified to be low enough. On the other hand, the filtering of the transposed quantization noise spectrum remains necessary on the output of the modulator.

Second, in order to restore a digital image of the signal as close as possible to the analog signal, it is necessary to resample the output signal of the modulator digitized at the frequency f_e/N, at least twice over the limit of the useful band f_0 by performing the reverse operation of the oversampling (from the bottom to the top in Figure 1.20). It is also important that it be filtered in order to eliminate the satellites of the signal spectrum as well as the out-of-band components originating from the system itself, which generates its own noise in addition to the noise due to the quantization error.

Third, a digital signal has to be generated encoded with an effective number of bits n (or resolution) of the same order as the number of equivalent bits, which corresponds to the theoretical signal-to-noise ratio of the modulator. Nonetheless, the number of bits m on the modulator output is much smaller than n. This operation requires to compute the sum (equivalent to the average apart from a constant) of 2^{n-m} successive samples encoded onto an m-bit datastream in order to achieve the resolution of n bits.

Since finite impulse response (FIR) filters are able to perform all of these tasks, we could just in principle include only one single filter. However, the required attenuation and especially the direct computation of the sum using the final resolution would require a prohibitive number of coefficients, much larger than that shown in Figure 1.56. As a matter of fact, for instance for a resolution of 20 bits, representing $2^{20} = 1.048 \times 10^6$ levels, there should be as many coefficients as the number of levels. It is therefore essential to perform the filtering in several stages, which also brings other benefits. Filtering the noise and increasing the resolution while at the same time lowering the frequency are part of a first stage with a transmittance H_1. Filtering intended to ensure the validity of the sampling theorem after decimation is performed in a second stage at the final sampling frequency with a transmittance H_2.

It remains nevertheless tantamount to use one or more FIR filters until the number n of bits required is not reached, because FIR filters have the intrinsic property of increasing the resolution at each operation without another digital function having to be implemented. The reason behind is that the output signal is computed as a sum of successive samples weighted by coefficients. In addition, these are filters of the "pipeline" type, meaning that they output a sample at the same rate as the input, namely at the frequency f_e in the case of the first filter, the steady state being reached as soon as the number of periods exceeds the number of filter coefficients. Therefore, there is a very wide latitude to set a sampling frequency on the output between f_e and f_e/N. To save the number of coefficients and avoid multiplication operations as well as the corresponding circuitry, we can resort to the averaging filter or comb filter (or even rectangular sampled impulse response filter) whose frequency response can be found in Figure 1.55. It is rather inefficient in terms of attenuation slope because the envelope of its frequency response beyond the cutoff frequency is that of a first order filter (–20dB/decade), in light of its transmittance (section 1.4.6.5) computed in the case of M coefficients:

$$H_{comb}(f) = \frac{1-\exp(-j2\pi(M+1)fT_e)}{1-\exp(-j2\pi fT_e)} = \frac{\sin(\pi(M+1)fT_e)}{\sin(\pi fT_e)}\exp(-jM\pi fT_e)$$

that crosses 0 for all multiple frequencies of $\dfrac{1}{(M+1)T_e}$ except $f_e = 1/T_e$ and its multiples. However, we can associate multiple filters of this type in cascade, three for example, with sampling frequency ratios between input and output, namely N_1 for the first, N_2 for the second, N_3 for the third, and such that $N = N_1\ N_2\ N_3$. The numbers of coefficients M_1, M_2, M_3 will be respectively greater than N_1, N_2, N_3 in order to give each filter the time to compute a non-truncated average value. The number of possible levels after the successive average of M_1, M_2 and M_3 samples encoded onto m bits will be $2^m M_1$ then $2^m\ M_1 M_2$ then $2^m M_1 M_2\ M_3$. In total, $\log_2\ (2^m M_1 M_2\ M_3)$ will have to slightly exceed the number of bits n desired for the final output. It will thus be possible to obtain transmittance zeros, respectively at frequencies $f_e\ /(M_1{+}1)$, $f_e\ /(N_1\ (M_2{+}1))$, $f_e\ /(N_1\ N_2\ (M_3{+}1))$ arranged in such fashion to be interleaved over the frequency scale. These transmittance zeros will be favorable to the increase in attenuation compared to the asymptotic value of −60dB/decade. Taking as an example $m = 1$ bit, $M_1 = 145$, $M_2 = 125$, $M_3 = 105$, which give $\log_2(2M_1M_2M_3) = 21.86$ for a final 20-bit resolution, and assuming a second-order modulator which imposes an oversampling rate R of about 400 for this resolution (Figure 2.28), we will take $N_1 = 5$, $N_2 = 8$ and $N_3 = 10$, that is $N = 400$.

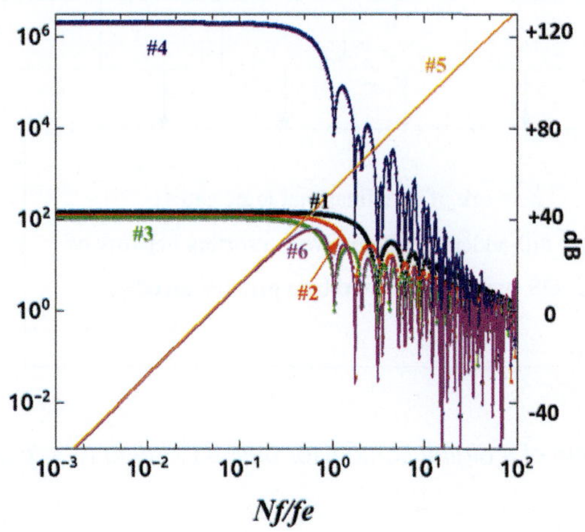

Figure 2.32. *Modulus of transmittances H_{comb1} (#1), H_{comb2} (#2), H_{comb3} (#3), H_1 which is the product of the previous three (#4), unfiltered (#5) and filtered noise voltage of the second-order modulator (#6) by the product of the three transmittances depending on the frequency normalized by the frequency at the end of the digital filtering chain f_e/N. For a color version of this figure, see www.iste.co.uk/muret/electronics3.zip.*

The responses $\left|H_{comb1}(f)\right| = \left|\frac{\sin(\pi(M_1+1)f/f_e))}{\sin(\pi f/f_e)}\right|$, $\left|H_{comb2}(f)\right| = \left|\frac{\sin(\pi(M_2+1)f/(N_1T_e))}{\sin(\pi f/(N_1f_e))}\right|$ and $\left|H_{comb3}(f)\right| = \left|\frac{\sin(\pi(M_3+1)f/(N_1N_2f_e))}{\sin(\pi f/(N_1N_2f_e))}\right|$ of the three comb filters according to the frequency normalized by the final sampling frequency f_e/N, that is to say the dimensionless variable fNT_e, and the product $\left|H_1\right| = \left|H_{comb1}(f)H_{comb2}(f)H_{comb3}(f)\right|$ are plotted in Figure 2.32, as well as the unfiltered and filtered noise voltages.

It turns out that the filtered noise voltage remains lower by at least 80 dB compared to the maximal transmittance, and with an even greater gap as we move toward the low frequencies. A more accurate evaluation of the signal-to-noise ratio would require to compute the integral of the filtered noise power on the entire frequency domain. However, it can be stated that the increase of the initial noise voltage (unfiltered) beyond f_e/N at the rate of +40dB/decade is replaced by a slight decrease, which demonstrates that the goal looked for is reached. We can generalize this method by adopting a number of comb filters greater by one unity to the order of the modulator.

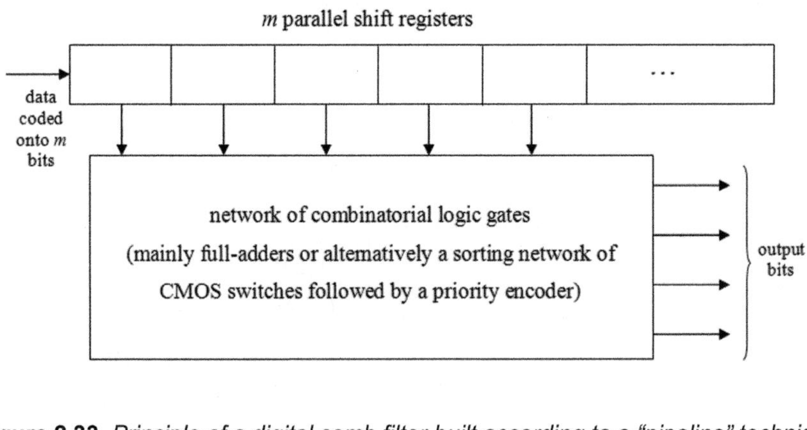

Figure 2.33. *Principle of a digital comb filter built according to a "pipeline" technique*

A comb filter structure is less complex than that of general FIR filters because all coefficients being the same (assumed unity), multiplication circuits are unnecessary. The most simple principle of such a digital filter is based on counters but outputs result only after adding all M_1, or M_2 or M_3 samples, according to a rhythm f_e/M_1, f_e/M_2, f_e/M_3, which is too low in comparison to frequency f_e/N_1 or f_e/N_2 or f_e/N_3, respectively. So, a system capable of delivering the result at higher throughput consists, for example, of a "pipeline" structure sketched in Figure 2.33. It is mainly built with shift registers and a network of combinatorial logic gates capable of performing additions of multi-bit words, which can be implemented with the help of methods of logical synthesis (see [NDJ 16a, NDJ 16b]). The previous numerical values of M_1, M_2, M_3, N_1, N_2, N_3 may be changed in order to comply with the requirements of the logical system.

The second filtering stage consists primarily in strengthening the attenuation beyond the frequency f_e /N, which was already present with a slope of −60 dB/decade, with an even steeper slope. Nonetheless, as in the case of the analog interpolators (see Chapter 1), the decrease of the modulus of transmittance H_1 of the first stage below the cut-off frequency f_e /N (Figure 2.34) must be compensated for if it is desirable to achieve a constant transmittance in the bandwidth as expected for an ideal low-pass filter; this requirement leads to a modulus of transmittance H_2 such as that of Figure 2.34.

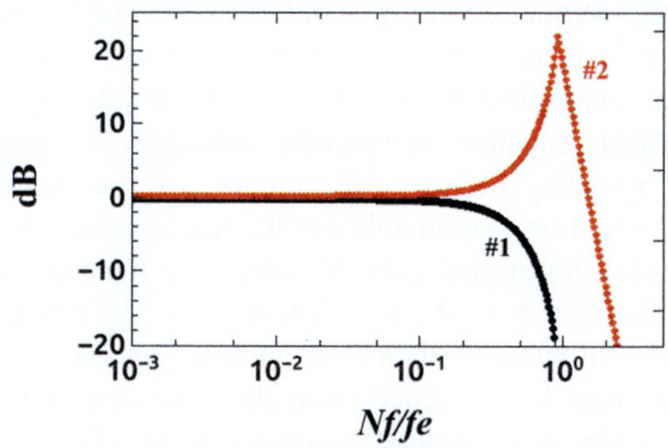

Figure 2.34. *Modulus of transmittances H_1 (#1) and H_2 (#2). For a color version of this figure, see www.iste.co.uk/muret/electronics3.zip*

In the event we want to keep the benefit provided by a linear phase shift with the frequency, which has the advantage of generating minimal distortion in impulse signals (see Chapter 5) and steep signals, a FIR-type filter is necessary. To obtain a modulus whose profile is indicated as in Figure 2.33, a very substantial number of coefficients is needed and, as a result, the circuitry capable of performing addition and multiplication operations which are necessary, attains a considerable volume. This is why manufacturers of integrated circuits have developed digital signal processors (DSP) that sequentially execute operations using the same arithmetic and logical unit and accumulate partial results in peripheral RAM memory according to an algorithm stored in ROM memory. This method is much less costly in terms of circuitry size and is limited only by the throughput rate of samples, which at this point is the slowest of the system because the decimation operation has already been made by the first filtering stage. It is thus possible to implement FIR filters making use of 256 coefficients which can be calculated by means of one of the synthesis methods indicated in the previous chapter and capable of approaching a template such as that defined by H_2 in the frequency domain.

2.5.6. *"Sigma-delta" DAC*

In principle, to obtain a DAC from a "sigma-delta" ADC, it suffices to replace analog functions by digital functions and vice versa. The converted signal is then extracted from the analog output. Theoretical developments achieved in DACs remain entirely valid because they apply to sampled systems, regardless of the analog or digital nature of signals. However, the transformation of analog functions of the modulator into digital functions is not straightforward because it implies changes in signal-processing technology and models. The overall function of the modulator actually becomes that of a section responsible for the transposition of the resolution and the rate of incoming digital data. In effect, the most appropriate way to carry out the conversion consists, in a first step, in converting n-bit encoded binary data into m-bit encoded data at a much higher rate and with $m << n$. This function can thus be separated from the final conversion stage which amounts to analog filtering preceded by a conventional DAC if $m > 1$.

2.5.6.1. *Resolution and digital rate conversions*

The M binary words representing data that arrive at frequency $f_1 = 1/T_1$ are encoded onto n bits, which corresponds to a decimal representation $M = \sum_{k=0}^{n-1} 2^k \, b_k\big|_{T_1}$ where the notation used for bits $b_k\big|_{T_1}$ indicates that they are available on the input of the modulator for a period T_1. The simplest case is considered first where $m = 1$ bit. Similar to what happens in the 1-bit analog-digital sigma-delta modulator, the desired goal is to transform the data encoding in order to obtain the data coded with 1 bit, that is, a value $d = 0$ or $d = 1$ for the output data. This operation generates a loss of resolution, which is accompanied by a huge relative quantification error since the binary value of the output word becomes $M' = d$. It must then be compensated by a much higher throughput rate of data d, at the frequency $f_2 = 1/T_2$, in order to be able to recover an average value of M' equal to M after a period of the same order as T_1 and to transfer the quantization noise spectral density to frequencies neighboring f_2. To this purpose, it is necessary to implement a negative feedback loop capable of canceling the error resulting from the difference between M and the average value of M'. It is therefore necessary to calculate this average value of M' or that of the difference $M - M'$ inside the loop. This is a function provided by an integrator in analog circuits that is replaced here by the digital circuit of an adder-subtractor (Figure 2.35), powered by the error signal delayed in the feedback loop. The transfer function thus implemented for the signal is identical to that of the integrator of the first-order modulator in the sigma-delta ADC with $m = 1$ bit. If this average value should be calculated over the total time allowing a variation of M equal to the full scale, more specifically $2^n - 1$, this would yield for a given frequency f_1 on input excessively high values of frequency f_2 inside the loop and on output. For example, for $n = 16$ bit and 44.1 kHz, the value utilized in compact discs, we would have $f_2 = 2.9$ GHz! It is admitted that the loop error be canceled only after several periods instead of a single one, during large amplitude variations of M. However, this state represents only a tiny fraction of the operating time because as stated at the beginning of the presentation of the sigma-delta system, its very principle consists of updating the output of the modulator loop such that the signal error for every clock tick is minimized. In the case $m = 1$ bit, the error signal d is in principle a bit of weight 2^{n-1}, but this weight can be chosen to be lower for the purpose of canceling out the loop error in a more gradual manner.

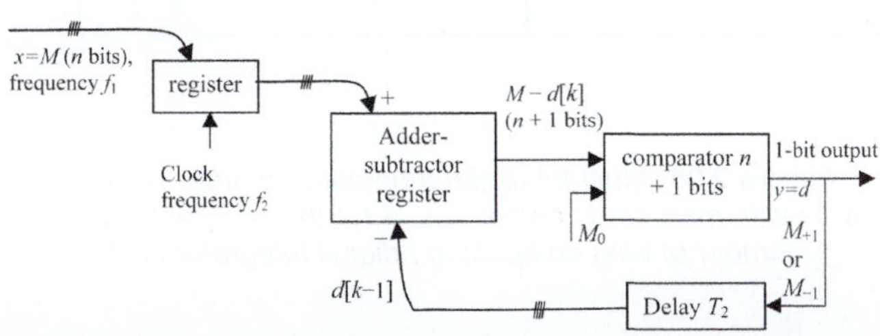

Figure 2.35. *Digital 1-bit modulator; in the chosen representation of signed numbers (using two's complement or half-scale shift) M_0 represents zero, $d = M_{+1}$ the positive number of weight 2^{n-1}, $d = M_{-1}$ the negative number of weight 2^{n-1}.*

It can be noticed in this first-order modulator that signal d, whose average value contains useful information, also contains noise since it is generated in the loop just before the output. It, however, undergoes the same processing as in analog modulators, namely a differentiation, insofar as the adder-subtractor register replaces the integrator by memorizing the delayed correction generated by the comparator for each clock period $T_2 = 1/f_2$. The noise transfer function is therefore the same as in the analog modulator of the first-order sigma-delta ADC. However, the practical limitation of the ratio of frequencies f_2 / f_1 to a value much smaller than 2^n can be overcome by solutions discussed here below.

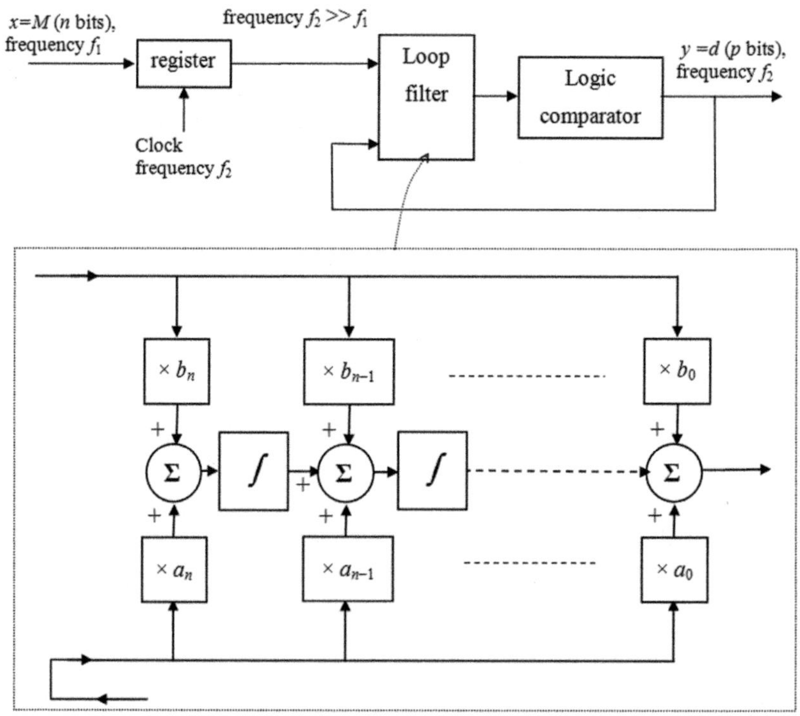

Figure 2.36. *Multi-bit digital modulator of order higher than 1 (with $m-n$ zero coefficients a_j for the compatibility of the number of bits) comprising l digital integrators ($\int$)*

The most effective solutions which allow going beyond the case of the digital 1-bit modulator consist in generating on output and in the loop a signal d digitized onto m bits, with $m > 1$, so as to reduce the ratio $2^n/2^m$, and to increase the order of the modulator to a value $l > 1$. In this case, we develop a modulator in which signals are processed by an lth-order loop filter (Figure 2.36) in order to benefit from the same advantages already studied in analog modulators and which is studied as a digital filter (see exercise 2.6.6). The error signal is then integrated several times and the signal-to-noise ratio is significantly improved, as can be noted from the formula shown for the order-l modulator in the DAC. Nonetheless, the stability of such a filter has to be verified (see exercise 2.6.6). Since it may be problematic for $l \geq 4$, it is in our interest to improve the signal-to-noise ratio by $6m$ dB with $m > 1$. For example, a gain of 30 dB is achieved with $m = 5$ bits. The logic comparator can be avoided by manipulating signed numbers, because the comparison with the number 0 (zero error) has to be executed on the filter output, which implies a subtraction of the number 0, equivalent to a nonexisting operation. Therefore, only the m bits with the strongest weights are the ones to be extracted at the filter output and which constitute the error signal fed back into the feedback loop. The MASH-type or interpolative-type modulator can also be employed in their digital form. The applications thereof include the storage of digital sound in the form encoded onto 1 bit into specific media (SACD). However, the main application is that of the sigma-delta DAC, described subsequently.

2.5.6.2. *Digital-to-analog sigma-delta conversion*

The output signal of the previous modulators is of the all-or-nothing type if the output encoding is carried out onto 1 single bit, and in this case its average value contains the analog information. This signal is a continuous flow of data of the same weight and for this reason it is called “bitstream”. The main characteristic of the 1-bit encoded signal is that the number of pulses in one of the two logical states per time unit determines its average value, and for this reason it is often referred to as pulse-density modulation signal. It can also be transformed into a PWM signal, or rather with modulated duty cycle using a counting technique. Conversely, if it is a m-bit encoded multi-bit signal, it is necessary to add a DAC. In this case, we will choose a DAC among those that convert all bits simultaneously, such as current weighted-based (section 2.3.1) or R-2R ladder converters (section 2.3.3) if the purpose is to utilize resistive technology, or one based on parallel redistribution of charge (section 2.3.4.2) if willing to use capacitive technology. In the first or the third case, extreme currents or capacitances are in the ratio 2^m but this does not significantly degrade the accuracy as m does practically not exceed 6 in reality, which corresponds to a maximal ratio of 64 between extreme resistances or capacitances.

The final signal can then be developed using analog filtering showing the same frequency response as the final filter previously studied for the ADC, that is to say implementing a low-pass transmittance with a sufficiently steep cutoff, to approach the ideal low-pass filter in order to reduce at best the quantization noise power, and to ensure the validity of the sampling theorem on the signal after having undergone decimation. This system is universally used in CD-ROM drives and other media for the transformation of digital sound encoding for its reproduction in analog form and probably would find many other applications due to its high level of performance and ease to be implemented as an integrated circuit [JAN 11].

2.6. Exercises

2.6.1. *DAC based on R-2R network and current sources*

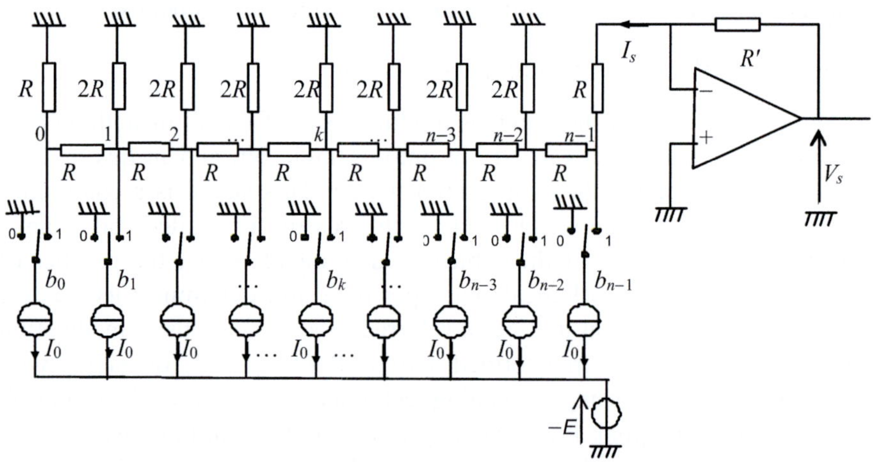

The above schematic is that of an n-bit R-2R network converter where nodes are numbered from 0 to $n-1$. Every combination of a switch plus a current source is actually a bipolar transistor operating in conductor state nonetheless unsaturated (current source I_0 when the control bit is 1) or switched off when the control bit is 0.

1) a) When $b_{n-1} = 1$ and all others 0, determine I_s according to I_0.

b) When $b_{n-2} = 1$ and all others 0, determine I_s according to I_0.

c) When $b_{n-3} = 1$ and all others 0, determine I_s according to I_0.

d) When $b_k = 1$ and all others 0, generalize by establishing the division of current I_0 at node k, then the division of current at node $k+1$ and so forth. Thereof deduce I_s as a function of $k - n$ and I_0.

2) Using the superposition theorem, give I_s in the general case based on I_0 and M where M is the decimal equivalent of the natural binary number b_{n-1} b_{n-2} b_{n-3} ... b_1 b_0 (digits placed in the order of decreasing binary weights). What is the value of R' which allows us to get $V_S = 2^{-n} R\, I_0 M$?

Answer:

It is important to notice that when switches are in position 0 (currents I_0 are then derived to ground and not in the R-2R network), the equivalent resistance of the left-hand side of any node k, without taking into account the resistance connected between this node k and the ground, is always equal to $2R$.

Let I_{k-1} be the current arriving from the left-hand side through this equivalent resistance from node $k-1$. Let I'_k be the current arriving at node k directly through the resistance $2R$ connected between this node and the ground. Therefrom, by current division we deduce the following.

1) a) Since I_s flows through R whereas I_{n-2} flows through $2R$: $I_s = 2I_{n-2}$ and $I_0 = I_s + I_{n-2}$. Hence:

$$I_s = \frac{2I_0}{3}.$$

b) On the right side of node $n-2$, we have a resistance $R+R = 2R$. Therefore, three $2R$ resistors in parallel traversed by the same currents: $I_s = I_{n-3} = I'_{n-2}$ and $I_0 = I_s + I'_{n-2} + I_{n-3}$. Hence:

$$I_s = \frac{I_0}{3}.$$

c) On the right-hand side of node $n-3$, we see a resistance $2R$. Therefore, we have again three $2R$ resistances in parallel traversed by the same currents.

However, the current I'_{n-3} is divided into two parts equal to I_s at node $n-2$. Thus, $2I_s = I_{n-4} = I'_{n-3}$ and $I_0 = 2I_s + I'_{n-3} + I_{n-4}$.

From which $I_s = \dfrac{I_0}{6}$.

d) By repeating the previous reasoning and noticing that the current arriving from the right-hand side of node k is equal to $2^{n-2-k} I_s$, we deduce that

$$I_s = \frac{4}{3} \times \frac{I_0}{2^{n-k}}.$$

2) Using the superposition theorem: $I_s = \dfrac{4}{3} I_0 \displaystyle\sum_{k=0}^{n-1} \dfrac{b_k}{2^{n-k}}$,

and in other words by establishing $M = \displaystyle\sum_{k=0}^{n-1} b_k\, 2^k$, we get: $I_s = \dfrac{4}{3} \dfrac{M\, I_0}{2^n}$.

Since $V_s = R'I_s = 2^{-n} R\, M\, I_0$, it is deduced that: $R' = \dfrac{3}{4} R$.

2.6.2. *Series DACs based on redistribution of charge*

1) Basic assembly

Binary variables h_0, h_1, h_2, b_i ($i = 0$ to $n-1$) assume the values 0 or 1 in the following circuit ($\overline{b_i}$ is the logical complement of b_i). The switches K_0, K_1, K'_1 and K_2 are respectively switched on (closed contact) when the associated binary variable specified in parentheses is set to 1, or open when set to the value 0, as indicated in the diagram:

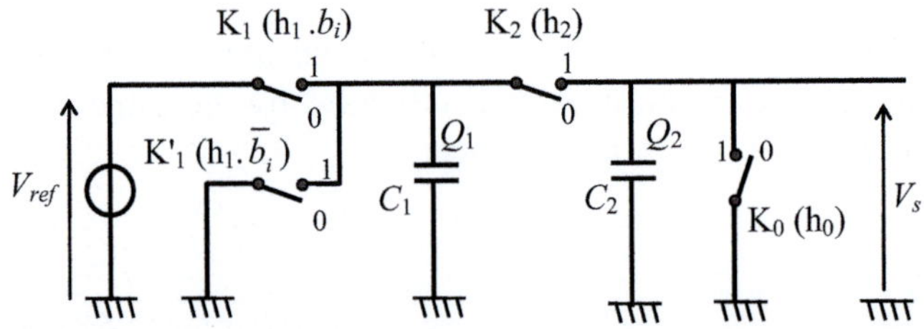

The two capacitances are identical: $C_1 = C_2 = C$. The sequence of the binary values of the switches control variables is as follows:

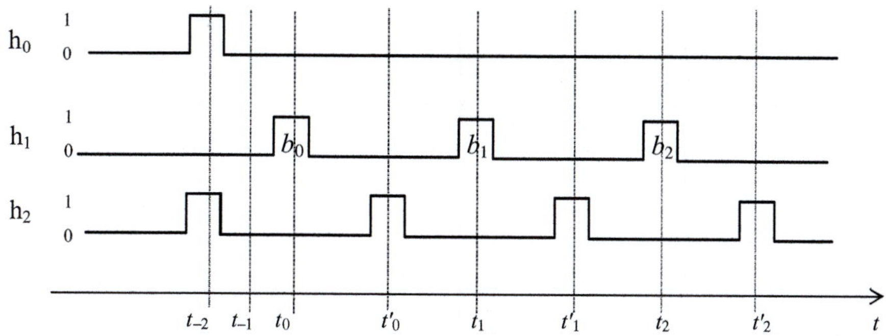

During the time intervals where h_1 is 1, bits b_0, b_1, b_2 and so on assume value 0 or 1 according to the numerical value of the corresponding binary word, b_0 being the LSB.

a) Phase t_{-2}: indicate the position of the switches and deduce therefrom $V_s(t_{-2})$, $Q_1(t_{-2})$ and $Q_2(t_{-2})$.

b) Phase t_{-1}: indicate the position of the switches and deduce therefrom $V_s(t_{-1})$, $Q_1(t_{-1})$ and $Q_2(t_{-1})$.

c) Phase t_0: taking into account the position of the switches, determine $Q_1(t_0)$ according to b_0, C, V_{ref}. What is the value of $Q_2(t_0)$?

d) Phase t'_0: taking into account the conservation of charge in the isolated portion of the circuit during the transition $t_0 \rightarrow t'_0$, determine $Q_1(t'_0)$, $Q_2(t'_0)$ and $V_s(t'_0)$ according to b_0, C, V_{ref}.

e) Phase t_1: taking into consideration the position of the switches, determine $Q_1(t_1)$ according to b_1, C, V_{ref}. What is the value of $Q_2(t_1)$?

f) Phase t'_1: taking into account the conservation of charge in the isolated portion of the circuit during the transition $t_1 \rightarrow t'_1$, determine $Q_1(t'_1)$, $Q_2(t'_1)$ and $V_s(t'_1)$ according to b_0, C, V_{ref}.

g) Using induction reasoning, determine $V_s(t'_{n-1})$ as a function of b_0, b_1, b_2, … b_{n-1} and V_{ref}.

h) Summarize the operation by completing the following table:

Time t	$Q_1(t)$	$Q_2(t)$	$V_s(t)$
t_{-2}			
t_{-1}			
t_0			
t'_0			
t_1			
t'_1			
t_2			
t'_2			

2) Operational amplifier assembly

The sequence of binary values of the control variables of the switches in the assembly of Figure 2.11 is the following:

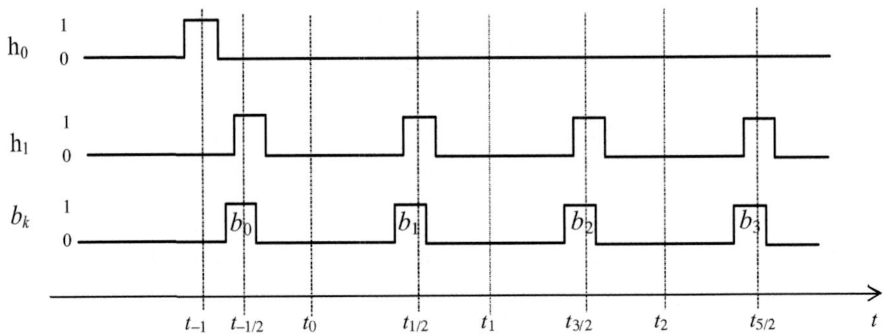

Charges Q_1 and Q_2, respectively, stored in capacitors C_1 and C_2 are those of the armatures connected to the minus input of the operational amplifier. The capacitances are equal: $C_1 = C_2 = C$.

a) Phase t_{-1}: redraw the diagram by replacing switches turned on with linking conductors and switches turned off with no connection at all. Therefrom, deduce the function performed by the operational amplifier which is assumed as being ideal (zero input current), except for the offset voltage V_{off} mentioned in the diagram. What is the value of charge $Q_2(t_{-1})$?

b) Phase $t_{-1/2}$: redraw the diagram by replacing the switches turned on by linking conductors and the switches turned off with no connection at all. Therefrom, deduce the function performed by the operational amplifier assumed to be ideal, except the offset voltage V_{off} mentioned in the diagram, as well as the charge $Q_1(t_{-1/2})$ according to C, V_{off} b_0 and V_{ref}. What is the value of charge $Q_2(t_{-1/2})$?

c) Phase t_0: redraw the diagram by replacing the switches turned on by linking conductors and the switches turned off with no connection at all. How are the two capacitors now connected? In the presence of a null input current in the operational amplifier, what happens to the charge of the capacitors? Considering the isolated system constituted by the armatures of the two capacitors connected to the minus input of the operational amplifier, establish the relation between charges at times $t_{-1/2}$ and t_0. Deduce the output voltage $V_s(t_0)$.

d) Phases $t_{1/2}$ and t_1: following the analogy with the two previous phases, deduce the expressions of $Q_1(t_{1/2})$ and $Q_2(t_{1/2})$, then those of $Q_1(t_1)$, $Q_2(t_1)$ and $V_s(t_1)$.

e) Following phases: using mathematical inductive reasoning, determine the result $V_s(t_{n-1})$ at the end of the conversion.

Answer:

1) Basic assembly

a) According to the values of the logical variables (in parentheses), K_0 and K_2 are closed (switched on), while K_1 and K'_1 are open (switched off). From which we can deduce that $Q_1(t_{-2}) = Q_2(t_{-2}) = 0$; $V_s(t_{-2}) = 0$.

b) All switches are open (turned off) and since capacitors are isolated, they retain a null charge. From which we can deduce that $V_s(t_{-1}) = 0$.

c) Capacitor C_1 is charged when subjected to voltage b_0V_{ref}, or more precisely $Q_1(t_0) = C\, b_0V_{ref}$ while the charge of C_2 remains zero: $Q_2(t_0) = 0$ because K_0 and K_2 remain open.

d) Both capacitors are arranged in parallel and their upper armatures form an isolated system. Therefore, $Q_1(t_0) + Q_2(t_0) = Q_1(t'_0) + Q_2(t'_0)$, that is $C\, b_0V_{ref} = 2\, Q_2(t'_0)$ because the capacitances are the same and so is the voltage; from which we can deduce that $V_s(t'_0) = b_0V_{ref}/2$.

e) K_2 having been switched off (open), $Q_2(t_1) = Q_2(t'_0)$ while C_1 is charged under $b_1 V_{ref}$, namely $Q_1(t_1) = C\, b_1 V_{ref}$.

f) Both capacitors are arranged in parallel and their upper armatures form an isolated system. Thereby, $Q_1(t_1) + Q_2(t_1) = Q_1(t'_1) + Q_2(t'_1) = 2\, Q_1(t'_1) = 2\, Q_2(t'_1) = 2\ C\ V_s(t'_1)$. By carrying forward, $C\, b_1 V_{ref} + C\ b_0 V_{ref}\ /2 = 2\, Q_2(t'_1) = 2\, C\, V_s(t'_1)$, that is $V_s(t'_1) = b_1 \frac{V_{ref}}{2} + b_0 \frac{V_{ref}}{4}$.

g) The repetition of the last two processes shows that the last voltage introduced into the terminals of C_1 is divided by 2 and added to previous ones themselves divided by 2 as many times as the number of repetitions; more specifically:

$$V_s(t'_{n-1}) = V_{ref}\left(\frac{b_{n-1}}{2} + \frac{b_{n-2}}{4} + \frac{b_{n-3}}{8} + \cdots + \frac{b_{n-k}}{2^k} + \cdots + \frac{b_0}{2^n}\right) =$$

$$\frac{V_{ref}}{2^n}\left(b_{n-1} 2^{n-1} + b_{n-2} 2^{n-2} + \cdots + b_{n-k} 2^{n-k} + \cdots + b_0\right).$$

h)

Time t	$Q_1(t)$	$Q_2(t)$	$V_s(t)$
t_{-2}	0	0	0
t_{-1}	0	0	0
t_0	$C\, b_0 V_{ref}$	0	0
t'_0	$C\, b_0 V_{ref}/2$	$C\, b_0 V_{ref}/2$	$b_0 V_{ref}/2$
t_1	$C\, b_1 V_{ref}$	$C\, b_0 V_{ref}/2$	$b_0 V_{ref}/2$
t'_1	$C\, b_1 V_{ref}/2 + C\, b_0 V_{ref}/4$	$C\, b_1 V_{ref}/2 + C\, b_0 V_{ref}/4$	$V_{ref}(b_1/2 + b_0/4)$
t_2	$C\, b_2 V_{ref}$	$C\, b_1 V_{ref}/2 + C\, b_0 V_{ref}/4$	$V_{ref}(b_1/2 + b_0/4)$
t'_2	$C\, V_{ref}(b_2/2 + b_1/4 + b_0/8)$	$C\, V_{ref}(b_2/2 + b_1/4 + b_0/8)$	$V_{ref}(b_2/2 + b_1/4 + b_0/8)$

2) Operational amplifier circuit

a) The capacitance C_1 is short-circuited and the operational amplifier is in voltage follower configuration, imposing the voltage V_{off} onto the armature of C_2 connected to the minus input. The charge of C_2 is thus $Q_2(t_{-1}) = C\ V_{off}$.

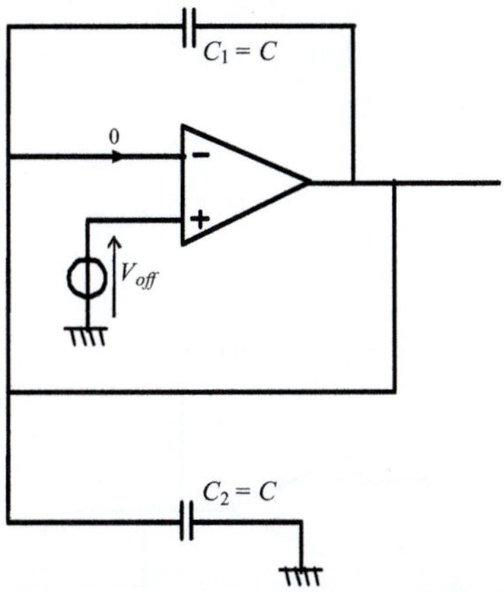

b) The operational amplifier is still in follower arrangement, imposing again the same voltage V_{off} on the armature of C_2 connected to the minus input. Since C_2 is disconnected from the ground, it retains its earlier charge and we still have $Q_2(t_{-1/2}) = C\ V_{off}$. At the same time, the capacitor C_1 is subject to a potential difference $b_0V_{ref} - V_{off}$, which leads to $Q_1(t_{-1/2}) = C\ (V_{off} - b_0\ V_{ref})$ taking into account the evaluation of charges on the armatures connected to the minus input of the operational amplifier.

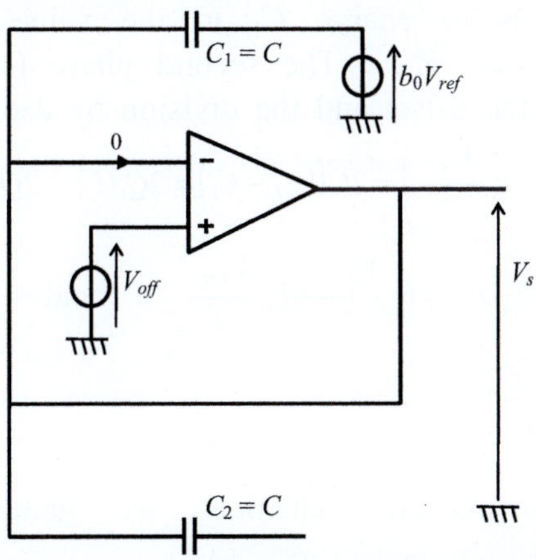

c) The two capacitors are connected in parallel and subjected to voltage $V_{off} - V_s$. Given that the input current of the operational amplifier is null, they preserve their total charge and at the same time provide the negative feedback that maintains a zero voltage between plus and minus inputs. However, since $C_1 = C_2 = C$ and because their previous charge was not the same, there is redistribution of charge: $Q_1(t_{-1/2}) + Q_2(t_{-1/2}) = Q_1(t_0) + Q_2(t_0) = 2\ Q_1(t_0) = 2\ Q_2(t_0)$, which leads to $C\ V_{off} + C\,(V_{off} - b_0\, V_{ref}\ ) = 2\,C\,(V_{off} - V_s)$, or by simplifying:

$$V_s(t_0) = b_0\,\frac{V_{ref}}{2} \text{ and } Q_2(t_0) = C\left(V_{off} - b_0\,\frac{V_{ref}}{2}\right).$$

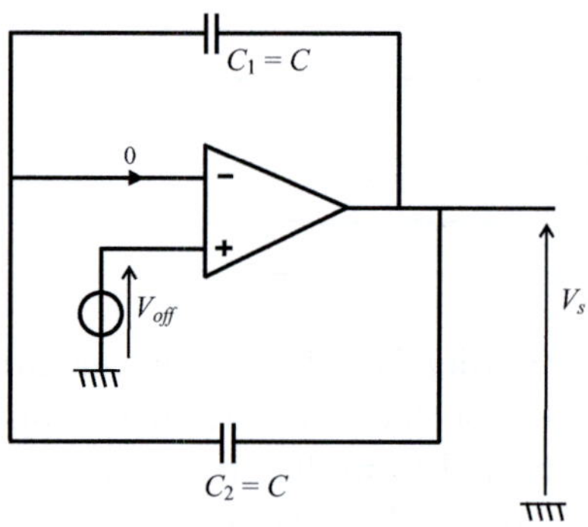

d) The conversion of the following bits, of increasingly stronger weights, is carried out by reiterating the two previous phases. The first one (at time $t_{1/2}$) keeps the previous charge of C_2, namely $Q_2(t_0)$, and allows us to charge C_1 to the value imposed by the next bit, namely $b_1\, V_{ref} - V_{off}$. The second phase (at time t_1) allows for the correction of the offset and the division by 2 of the total charge: $C\left(V_{off} - b_1 V_{ref}\right) + C\left(V_{off} - b_0\,\frac{V_{ref}}{2}\right) = 2C\left(V_{off} - V_s\right) = 2Q_1(t_1) = 2Q_2(t_1)$; from which we deduce: $V_s(t_1) = b_1\,\frac{V_{ref}}{2} + b_0\,\frac{V_{ref}}{4}$ and $Q_2(t_1) = Q_2(t_1) = C\left(V_{off} - b_1\,\frac{V_{ref}}{2} - b_0\,\frac{V_{ref}}{4}\right)$.

e) Using mathematical induction, we generalize: $V_s(t_{n-1}) = \frac{V_{ref}}{2}\left(b_{n-1} + \frac{b_{n-2}}{2} + \frac{b_{n-3}}{4} + \cdots + \frac{b_0}{2^{n-1}}\right) = \frac{V_{ref}}{2^n}\sum_{k=0}^{n-1} b_k\, 2^k$, the value obtained at the end of the conversion.

2.6.3. *Parallel DACs based on redistribution of charge and reduced capacitance*

The diagram of this converter is given in Figure 2.13 of the course.

a) In the first phase where $h_0 = 1$ (closed switch) and $h_1 = 0$ (open switch), at time t_0, write the equations of total charges on the upper armatures of the capacitors of part A equal to $Q_A(t_0)$, of part B equal to $Q_B(t_0)$, and charges stored in C_0 and C_2, according to V_1, V_{off} and b_k V_{ref}.

b) In the second phase where $h_0 = 0$ and $h_1 = 1$, at time t_1, write the equations of charges according to the voltages emerging at the various points of the circuit: $V_1(t_1) = V'_1$, V_{off} and V_s. Then establish the relation of the charge conservation in the two isolated systems A and B. Therefrom, deduce the output voltage $V_s(t_1)$.

Answer:

a) The operational amplifier is in follower configuration and the right armature of C_2 to the ground. Hence the charge of the left armature of C_2 is $Q_2(t_0) = C_2\, V_{off} = 2^{n/2}\, C\, V_{off}$. Defining $V_1(t_0) = V_1$, the whole charge in the upper armature of capacitors of part A is:

$$Q_A(t_0) = C\left(2^{\frac{n}{2}-1}(V_1 - b_{\frac{n}{2}-1}V_{ref}) + \cdots + 2(V_1 - b_1V_{ref}) + (V_1 - b_0V_{ref})\right)$$

$$= C\left(2^{\frac{n}{2}-1} + \cdots + 2 + 1\right)V_1 - C\left(2^{\frac{n}{2}-1}b_{\frac{n}{2}-1} + \cdots + 2b_1 + b_0\right)V_{ref} =$$

$$C\left(2^{\frac{n}{2}} - 1\right)V_1 - C\left(2^{\frac{n}{2}-1}b_{\frac{n}{2}-1} + \cdots + 2b_1 + b_0\right)V_{ref},$$

whereas the whole charge in the upper armature of capacitors of part B is

$$Q_B(t_0) = C\left(2^{\frac{n}{2}-1}(V_{off} - b_{n-1}V_{ref}) + \cdots + 2(V_{off} - b_{\frac{n}{2}+1}V_{ref}) + (V_{off} - b_{\frac{n}{2}}V_{ref})\right)$$

which by simplifying becomes

$$Q_B(t_0) = C\left(2^{\frac{n}{2}} - 1\right)V_{off} - C\left(2^{\frac{n}{2}-1}b_{n-1} + \cdots + 2b_{\frac{n}{2}+1} + b_{\frac{n}{2}}\right)V_{ref}.$$

The capacitor C_0 is charged under $V_1(t_0) - V_{off}$, namely a charge $C(V_1 - V_{off})$ which adds up to $Q_A(t_0)$ to form the total charge in the armatures of all capacitors in part A.

b) The new charges at time t_1 are written according to the voltages at different points of the circuit and to the values of capacitances. We establish $V_1(t_1)= V'_1$. For part A:

$$Q_A(t_1) = C\left(2^{\frac{n}{2}-1} + 2^{\frac{n}{2}-2} + \cdots + 2 + 1\right)V_1' = \left(2^{\frac{n}{2}} - 1\right)CV_1'$$

and $Q_B(t_1) = C\left(2^{\frac{n}{2}-1} + 2^{\frac{n}{2}-2} + \cdots + 2 + 1\right)V_{off} = \left(2^{\frac{n}{2}} - 1\right)CV_{off}$ for part B.

On the other hand, $Q_2(t_1) = 2^{n/2}\,C\,(V_{off} - V_s)$, and there is a charge on the right armature of C_0 equal to $Q_0(t_1) = C\,(V_{off} - V'_1)$.

All of the armatures of capacitors of part A along with the left armature of C_0 form a separate isolated system distinct from of all of the armatures of capacitors of part B, of the right armature of C_0 and of the left armature of C_2, which in turn form another isolated system. We should therefore separately address the conservation of charges in these two systems during transition $t_0 \rightarrow t_1$, which yields the two relations:

For part A: $Q_A(t_0) + C(V_1 - V_{off}) = Q_A(t_1) - Q_0(t_1)$

$$C\left(2^{\frac{n}{2}} - 1\right)V_1 - C\left(2^{\frac{n}{2}-1} b_{\frac{n}{2}-1} + \cdots + 2b_1 + b_0\right)V_{ref} + C(V_1 - V_{off}) = C\left(2^{\frac{n}{2}} - 1\right)CV_1' + C(V_1' - V_{off})$$

or by simplifying: $V_1' - V_1 = -2^{-\frac{n}{2}}\left(2^{\frac{n}{2}-1} b_{\frac{n}{2}-1} + \cdots + 2b_1 + b_0\right)V_{ref}$.

For part B: $Q_B(t_0) - C(V_1 - V_{off}) + Q_2(t_0) = Q_B(t_1) - C(V'_1 - V_{off}) + Q_2(t_1)$, that is

$$\left(2^{\frac{n}{2}}-1\right)V_{off}-\left(2^{\frac{n}{2}-1}b_{n-1}+\cdots+2b_{\frac{n}{2}+1}+b_{\frac{n}{2}}\right)V_{ref}+(V_{off}-V_1)+2^{\frac{n}{2}}V_{off}$$

$$=\left(2^{\frac{n}{2}}-1\right)V_{off}+(V_{off}-V_1')+2^{\frac{n}{2}}(V_{off}-V_s).$$

After carrying over and eliminating the difference $V'_1 - V_1$ from the previous result, we get:

$$V_s=2^{-\frac{n}{2}}\left(2^{\frac{n}{2}-1}b_{n-1}+\cdots+2b_{\frac{n}{2}+1}+b_{\frac{n}{2}}\right)V_{ref}+2^{-n}\left(2^{\frac{n}{2}-1}b_{\frac{n}{2}-1}+\cdots+2b_1+b_0\right)V_{ref}$$

$$=\frac{V_{ref}}{2^n}\left(2^{n-1}b_{n-1}+2^{n-2}b_{n-2}+\cdots+2^{\frac{n}{2}}b_{\frac{n}{2}}+2^{\frac{n}{2}-1}b_{\frac{n}{2}-1}+\cdots+2b_1+b_0\right)=\frac{V_{ref}}{2^n}\sum_{k=0}^{n-1}2^k b_k.$$

2.6.4. *Basic "delta-sigma" ADC*

Considering the following "delta-sigma" ADC, with V_a as analog input voltage and V_{ref} as perfectly constant reference voltage:

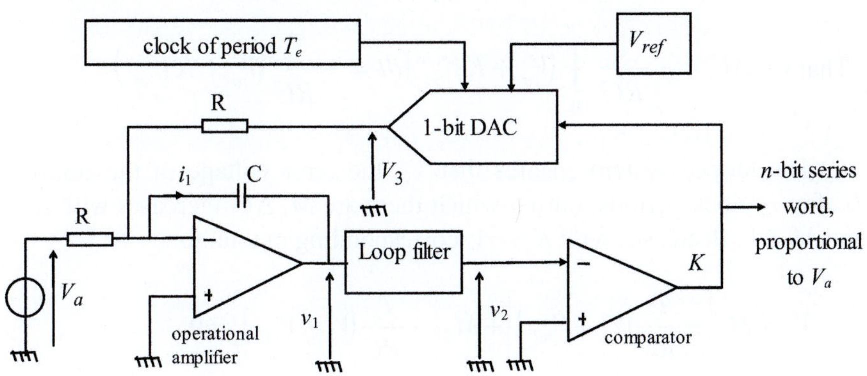

On the comparator output, K is an integer equal to +1 (or H level) if V_2 is negative and –1 (or L level) if V_2 is positive.

The voltage V_3 is equal to $K\ V_{ref}$ during each clock period T_e.

The operational amplifier is assumed to be ideal.

1) Calculate the change ΔV_1 of voltage V_1 obtained after a time interval equal to a clock period T_e, as a function of RC, T_e, V_a, K and V_{ref}.

2) The output voltage V_2 of the loop filter can be considered as the sum of N successive ΔV_1 increases. Of these N clock periods, there are M_1 during which $V_3 = +V_{ref}$ and M_2 during which $V_3 = -V_{ref}$, with $N = M_1 + M_2$. The feedback loop ensures that V_2, the error voltage of the control system, is 0 after these N clock periods. Deduce thereof the relation giving the ratio $\dfrac{V_a}{V_{ref}}$.

3) What should be the sign of V_{ref} for the error voltage V_2 to be actually canceled out (elaborate based on the signs of the various voltages)?

Answer:

1) Let i_1 be the current flowing through capacitor C toward the output of the operational amplifier:

$$i_1 = -C\frac{dv_1(t)}{dt} \Rightarrow \Delta V_1 = -\frac{1}{C}\int_0^{T_e} i_1 dt \text{ where } i_1 = \frac{1}{R}\left(V_a + KV_{ref}\right).$$

That is: $\Delta V_1 = -\dfrac{1}{RC}\displaystyle\int_0^{T_e}\left(V_a + KV_{ref}\right)dt = -\dfrac{T_e}{RC}\left(V_a + KV_{ref}\right)$

2) The looped system ensures that V_2, the error voltage of the control, is 0 after N clock periods, during which there are M_1 ΔV_1 increases with $K = 1$ and M_2 ΔV_1 decreases with $K = -1$, corresponding in total to

$$V_2 = M_1\left[-\frac{T_e}{RC}\left(V_a + V_{ref}\right)\right] + M_2\left[-\frac{T_e}{RC}\left(V_a - V_{ref}\right)\right] = 0.$$

Taking $N = M_1 + M_2$ into account, we deduce that $\boxed{V_a = V_{ref}\dfrac{M_2 - M_1}{N}}$ because T_e and RC disappear.

3) If $V_a > 0$, for example, and assuming that V_3 and V_1 are initially 0, V_1 and V_2 become negative (see question 1) $\Rightarrow K = +1 \Rightarrow V_3 = V_{ref}$ must be negative so as $i_1 = \frac{1}{R}\left(V_a + KV_{ref}\right)$ be negative in order to reverse the direction of the variation of V_2. Following a similar reasoning in the case where $V_a < 0$, we always arrive at the conclusion that V_{ref} must be negative.

2.6.5. *Third-order "MASH" modulator*

We study the "delta-sigma" ADC schematized in Figure 2.29, of the third-order "MASH" type.

1) Demonstrate the relation $S_1(z) = z^{-1} X(z) + (1 - z^{-1}) E_1(z)$. Deduce therefrom the expressions of the output signal of the integrator of the first stage then of $X_1(z)$.

2) Determine the equation giving $S_2(z)$, the output signal of the integrator of the second stage and finally $X_2(z)$.

3) Determine the equation giving $S_3(z)$.

4) Based on the previous relations, demonstrate that $Y(z) = z^{-1} X(z) + (1 - z^{-1})^3 E_3(z)$.

5) What multiplicative factor R of the sampling frequency should be chosen in order to achieve 12 equivalent resolution bits? 24 bits?

Answer:

1) On the output of the integrator of the first stage, we have $\left[X(z) - S_1(z)\right]\frac{z^{-1}}{1-z^{-1}}$ that has to be added to the noise $E_1(z)$ to obtain $S_1(z)$: $S_1(z) = \left[X(z) - S_1(z)\right]\frac{z^{-1}}{1-z^{-1}} + E_1(z)$. From which it is deduced that $S_1(z)\left[1+\frac{z^{-1}}{1-z^{-1}}\right] = S_1(z)\frac{1}{1-z^{-1}} = X(z)\frac{z^{-1}}{1-z^{-1}} + E_1(z)$, then $S_1(z) = z^{-1} X(z) + (1 - z^{-1}) E_1(z)$. It is then possible to recalculate $\left[X(z) - S_1(z)\right] \frac{z^{-1}}{1-z^{-1}} = \frac{z^{-1}X(z) - z^{-1}S_1(z)}{1-z^{-1}} = S_1(z) - E_1(z)$, which is a signal from which $S_1(z)$ is subtracted to yield: $X_1(z) = -E_1(z)$.

2) On the input of stage 2, the sum $X_1(z) + S_2(z)$ is achieved and integrated by means of an undelayed integrator and finally added to $E_2(z)$ to give $S_2(z)$: $S_2(z)=[X_1(z)+S_2(z)]\frac{1}{1-z^{-1}}+E_2(z)$; from which $S_2(z)\left[1-\frac{1}{1-z^{-1}}\right]=\frac{-z^{-1}S_2(z)}{1-z^{-1}}=X_1(z)\frac{1}{1-z^{-1}}+E_2(z)$ is deduced and can be rewritten as

$$z^{-1}S_2(z)=-X_1(z)-\left(1-z^{-1}\right)E_2(z)=E_1(z)-\left(1-z^{-1}\right)E_2(z).$$

Next, $X_2(z)$ is injected into the input of stage 3, obtained by subtracting $S_2(z)$ from the output of the integrator of stage 2:

$$X_2(z)=[X_1(z)+S_2(z)]\frac{1}{1-z^{-1}}-S_2(z)=-E_1(z)\frac{1}{1-z^{-1}}+$$
$$S_2(z)\left[\frac{1}{1-z^{-1}}-1\right]=-E_1(z)\frac{1}{1-z^{-1}}+\frac{z^{-1}S_2(z)}{1-z^{-1}},$$

that is finally:

$$X_2(z)=-\frac{E_1(z)}{1-z^{-1}}+\frac{E_1(z)-\left(1-z^{-1}\right)E_2(z)}{1-z^{-1}}=-E_2(z).$$

3) The integrator of stage 3 processes the signal $X_2(z) + S_3(z)$ and then the noise $E_3(z)$ is added to obtain $S_3(z)=[X_2(z)+S_3(z)]\frac{1}{1-z^{-1}}+E_3(z)$ or $S_3(z)\left[1-\frac{1}{1-z^{-1}}\right]=\frac{X_2(z)}{1-z^{-1}}+E_3(z)$. From which it can inferred that; $z^{-1}S_3(z)=-X_2(z)-\left(1-z^{-1}\right)E_3(z)=E_2(z)-\left(1-z^{-1}\right)E_3(z)$

4) The signals resulting from the action of the delays and differentiators on $S_2(z)$ and $S_3(z)$ are subtracted from $S_1(z)$ to form the signal $Y(z)$. The action of the first delays allows us to obtain $z^{-1}\,S_2(z)$ and $z^{-1}\,S_3(z)$, which are then respectively differentiated once and twice, thereby totaling:

$$Y(z)=S_1(z)-\left(1-z^{-1}\right)z^{-1}S_2(z)-\left(1-z^{-1}\right)^2z^{-1}S_3(z),$$

By carrying forward the previous results, it follows that

$$Y(z)=z^{-1}X(z)+\left(1-z^{-1}\right)E_1(z)-\left(1-z^{-1}\right)E_1(z)+\left(1-z^{-1}\right)^2E_2(z)-\left(1-z^{-1}\right)^2E_2(z)+\left(1-z^{-1}\right)^3E_3(z),$$

that is finally: $Y(z)=z^{-1}X(z)+\left(1-z^{-1}\right)^3E_3(z)$.

5) Based on the graph of Figure 2.28, a factor $R = 16$ is enough to obtain 12 equivalent bits and 256 have to be used for getting 24 equivalent bits.

2.6.6. *Third-order digital filter of a multi-bit "sigma-delta" DAC*

The block diagram of the filter is given in Figure 2.35.

1) Redraw the diagram of the filter with three undelayed digital integrator stages ($l = 3$) and assigning to the samples generated on output of each integrator the names $y_1[k]$, $y_2[k]$ and $y_3[k]$ in order from left to right. We call $x[k]$ the samples transmitted to the input (branch located on the top left of the diagram) and call $y[k]$ those obtained on output of the last adder and fed back into the loop through the branches located at the bottom of the diagram. It is not necessary to process the signal $y[k]$ using a logical comparator when the numbers are signed and compared to the number 0. The diagram will therefore include a direct feedbacking of $y[k]$ onto the multiplicative coefficients a_0, … a_3. The clock frequency is f_2.

2) Write the system of three recurrence equations that allow for expressing $y_1[k]$, $y_2[k]$ and $y_3[k]$ according to $y_1[k-1]$, $y_2[k-1]$, $y_3[k-1]$, $x[k]$ and $y[k]$ and to the relevant multiplicative coefficients in the diagram. In addition, write the equation verified by the sample $y[k]$ obtained on the output of the fourth adder. How can the order of the system be checked?

3) By substituting $y[k]$ by its expression taken from the last equation, determine the matrix equation $\mathbf{Y}[k-1] = \mathbf{M}\,\mathbf{Y}[k] + \mathbf{B'_0}\,x[k]$, where $\mathbf{Y}[k]$ and $\mathbf{Y}[k-1]$ are the column vectors respectively having as elements $y_1[k]$, $y_2[k]$, $y_3[k]$ and $y_1[k-1]$, $y_2[k-1]$, $y_3[k-1]$. What does the matrix **M** represent?

4) Determine the symbolic expression of the equation that enables us to calculate the poles z_1, z_2, z_3 of the system. Let r be a root of the characteristic equation. What are the conditions to be implemented in order for the system to be stable?

Since the main function of the filter is to repeat three times an integral of a linear combination of samples, we will consider studying the stability only in the particular case where the system exhibits a positive, real triple pole $z_1 = z_2 = z_3 = r^{-1}$. In this case, determine the condition on the parameter $a'_1 = \dfrac{a_1}{a_0 - 1}$ that enables us to ensure the stability of the system and the expressions of the parameters $a'_2 = \dfrac{a_2}{a_0 - 1}$ and $a'_3 = \dfrac{a_3}{a_0 - 1}$ according to a'_1.

5) Since the samples transmitted in the feedback loop are encoded onto a limited number of bits (typically ≤ 6), it should be considered that the ZT of the quantization noise $E_N(z)$ is added in the last adder. Determine the closed loop transfer function in the domain of the z variable for the signal, more specifically $H_S(z) = Y(z)/X(z)$ when $E_N(z) = 0$, and for the noise, namely $H_N(z) = Y(z)/E_N(z)$ when $X(z) = 0$.

Let us establish $b'_0 = \dfrac{b_0}{a_0 - 1}$; $b'_1 = \dfrac{b_1}{a_0 - 1}$; $b'_2 = \dfrac{b_2}{a_0 - 1}$; $b'_3 = \dfrac{b_3}{a_0 - 1}$; $E'_N(z) = \dfrac{E_N(z)}{a_0 - 1}$.

What are the conditions for the triple pole r^{-1} and coefficient a'_1 that allow us to approach a third-order derivative of the noise? When achieved, what is the consequence for the closed loop transmittance applied to the input signal $X(z)$?

Finally, compare this kind of filter to the interpolating filter of section 2.5.4 modified to be purely digital (suppressing ADC and DAC), and discuss advantages and drawbacks of each.

Answer:

1) The following functional diagram satisfies the conditions set out in question 1:

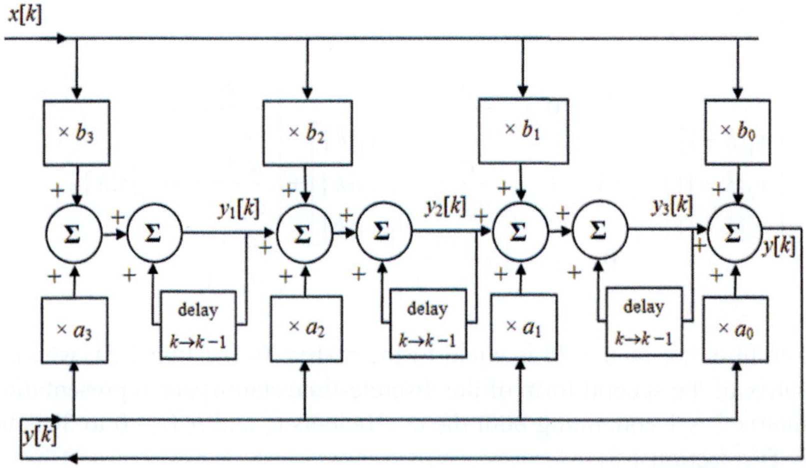

2) The integrators of the previous diagram output the following samples:

$$y_1[k] = y_1[k-1] + a_3\, y[k] + b_3\, x[k];$$
$$y_2[k] = y_2[k-1] + y_1[k] + a_2\, y[k] + b_2\, x[k];$$
$$y_3[k] = y_3[k-1] + y_2[k] + a_1\, y[k] + b_1\, x[k].$$

In addition, the samples present in the last adder verify:

$y[k] = a_0\ y[k] + y_3[k] + b_0\, x[k]$, which is re-written as $(1 - a_0)\, y[k] = y_3[k] + b_0\, x[k]$.

There are three distinct samples that can be found with a delay of one sampling period (index k–1); therefore, the system is of order 3.

3) By substituting $y[k]$ by its expression taken from the last equation in the system formed by the first three equations, we get:

$$\begin{bmatrix} 1 & 0 & \dfrac{a_3}{a_0-1} \\ -1 & 1 & \dfrac{a_2}{a_0-1} \\ 0 & -1 & 1+\dfrac{a_1}{a_0-1} \end{bmatrix} \begin{bmatrix} y_1[k] \\ y_2[k] \\ y_3[k] \end{bmatrix} = \begin{bmatrix} y_1[k-1] \\ y_2[k-1] \\ y_3[k-1] \end{bmatrix} + \begin{bmatrix} b_3 - \dfrac{a_3 b_0}{a_0-1} \\ b_2 - \dfrac{a_2 b_0}{a_0-1} \\ b_1 - \dfrac{a_1 b_0}{a_0-1} \end{bmatrix} \big[x[k]\big].$$

Or even

$$\begin{bmatrix} y_1[k-1] \\ y_2[k-1] \\ y_3[k-1] \end{bmatrix} = \begin{bmatrix} 1 & 0 & \dfrac{a_3}{a_0-1} \\ -1 & 1 & \dfrac{a_2}{a_0-1} \\ 0 & -1 & 1+\dfrac{a_1}{a_0-1} \end{bmatrix} \begin{bmatrix} y_1[k] \\ y_2[k] \\ y_3[k] \end{bmatrix} + \begin{bmatrix} \dfrac{a_3 b_0}{a_0-1}-b_3 \\ \dfrac{a_2 b_0}{a_0-1}-b_2 \\ \dfrac{a_1 b_0}{a_0-1}-b_1 \end{bmatrix} [x[k]].$$

Therefore, the matrix **M** is equal to the matrix **A'** (section 1.5), which is the matrix of the second form of the discrete-time state-space representation. The matrix 3 × 1 containing both the coefficients a_i and b_i ($i = 0$ to 3) is the matrix $\mathbf{B'_0}$ (section 1.5).

4) The poles are thus the inverse of the eigenvalues of **M**, which is the solution to the characteristic equation:

$$\det \begin{bmatrix} 1-\lambda & 0 & \dfrac{a_3}{a_0-1} \\ -1 & 1-\lambda & \dfrac{a_2}{a_0-1} \\ 0 & -1 & 1+\dfrac{a_1}{a_0-1}-\lambda \end{bmatrix} = 0,$$

which is written as $(1-\lambda)^2\left(1-\lambda+\dfrac{a_1}{a_0-1}\right)+\dfrac{a_3}{a_0-1}+(1-\lambda)\dfrac{a_2}{a_0-1}=0$;

or after expansion and by defining $a'_1=\dfrac{a_1}{a_0-1}$, $a'_2=\dfrac{a_2}{a_0-1}$, $a'_3=\dfrac{a_3}{a_0-1}$:

$$\lambda^3-(3+a'_1)\lambda^2+(3+2a'_1+a'_2)\lambda-(1+a'_1+a'_2+a'_3)=0.$$

The roots of this equation give the inverse of the poles z_1, z_2, z_3 of the system since **M** is the state matrix of the second form of the discrete-time state-space representation.

In order for the system to be stable, it is necessary that the three conditions $|z_1|<1$; $|z_2|<1$; $|z_3|<1$ be verified. Or equivalently, that the roots of the characteristic equation verify: $|\lambda_1| = \frac{1}{|z_1|} > 1$; $|\lambda_2| = \frac{1}{|z_2|} > 1$; $|\lambda_3| = \frac{1}{|z_3|} > 1$.

If there is a triple real root r, which will be assumed to be positive, this condition will be reduced to $r > 1$. Moreover, the characteristic equation is written as follows:

$$(\lambda - r)^3 = \lambda^3 - 3r\lambda^2 + 3r^2\lambda - r^3 = 0.$$

We deduce that $3r = 3 + a'_1$ or even $r = 1 + \frac{a'_1}{3} > 1$, implying that $a'_1 > 0$ to guarantee stability. Because this condition is relatively little constraining and easy to complete, we can presumably assume that it is possible to depart from it to obtain three real but distinct poles while continuing to guarantee the stability of the system. Nonetheless, it would be necessary to verify to what extent we can move parameters away from this condition by means of a computation of the numeric value of the poles with some software, and what benefits could be obtained from this development (see more details in question 5).

To obtain a triple root, the other two coefficients must satisfy additional requirements, namely $3r^2 = 3 + 2a'_1 + a'_2$ and $r^3 = 1 + a'_1 + a'_2 + a'_3$. From which it can be deduced that $a'_2 = \frac{a'^2_1}{3}$ and $a'_3 = \frac{a'^3_1}{27}$.

5) By adding the noise source $E_N(z)$ to the last adder, the ZT of the four equations of question 2 yields the system:

$$Y_1(z) = z^{-1}\, Y_1(z) + a_3\, Y(z) + b_3\, X(z);$$
$$Y_2(z) = z^{-1}\, Y_2(z) + Y_1(z) + a_2\, Y(z) + b_2\, X(z);$$
$$Y_3(z) = z^{-1}\, Y_3(z) + Y_2(z) + a_1\, Y(z) + b_1\, X(z);$$
$$(1 - a_0)\, Y(z) = Y_3(z) + b_0\, X(z) + E_N(z).$$

That is expressing $Y_3(z)$, $Y_2(z)$ and $Y_1(z)$ according to $Y(z)$ and carrying over onto the last equation:

$$(1-a_0)Y(z)=\frac{a_3Y(z)+b_3X(z)}{\left(1-z^{-1}\right)^3}+\frac{a_2Y(z)+b_2X(z)}{\left(1-z^{-1}\right)^2}+$$
$$\frac{a_1Y(z)+b_1X(z)}{1-z^{-1}}+b_0X(z)+E_N(z)$$

or

$$-Y(z)=\frac{a'_3\,Y(z)+b'_3\,X(z)}{\left(1-z^{-1}\right)^3}+\frac{a'_2\,Y(z)+b'_2\,X(z)}{\left(1-z^{-1}\right)^2}+$$
$$\frac{a'_1\,Y(z)+b'_1\,X(z)}{1-z^{-1}}+b'_0\,X(z)+E'_N(z)$$

or still

$$-\left[\left(1-z^{-1}\right)^3+a'_3+a'_2\left(1-z^{-1}\right)+a'_1\left(1-z^{-1}\right)^2\right]Y(z)$$
$$=\left[b'_3+b'_2\left(1-z^{-1}\right)+b'_1\left(1-z^{-1}\right)^2+b'_0\left(1-z^{-1}\right)^3\right]X(z)+\left(1-z^{-1}\right)^3E'_N(z)$$

which yields:

$$Y(z)=\frac{\left[-b'_0\,z^{-3}+(3b'_0+b'_1)z^{-2}-(3b'_0+2b'_1+b'_2)z^{-1}+(b'_0+b'_1+b'_2+b'_3)\right]X(z)+\left(1-z^{-1}\right)^3E'_N(z)}{z^{-3}-(3+a'_1)z^{-2}+(3+2a'_1+a'_2)z^{-1}-(1+a'_1+a'_2+a'_3)}.$$

The denominator is the same as the determinant of the matrix **M** = **A'**, and therefore can be replaced by $(z^{-1}-r)^3$.

The transmittance for the signal is thus

$$H_S(z)=\frac{\left[-b'_0\,z^{-3}+(3b'_0+b'_1)z^{-2}-(3b'_0+2b'_1+b'_2)z^{-1}+(b'_0+b'_1+b'_2+b'_3)\right]}{\left(z^{-1}-r\right)^3}$$

and for the noise: $H_N(z)=\dfrac{\left(1-z^{-1}\right)^3}{(a_0-1)\left(z^{-1}-r\right)^3}$.

In order to minimize noise at low frequencies and reject most of its spectrum toward higher frequencies, it is known that it proves beneficial to perform several derivations of this noise, and consequently to approach a transmittance $\frac{\left(1-z^{-1}\right)^3}{K}$ in which K is a constant. This is all the better achieved in $H_N(z)$ when r is large compared to 1, $r >> 1$, which is obtained with $a'_1 >> 1$. This also guarantees the stability.

In case a triple integration for the transmittance of the signal is desired, it would be preferable to approximate it by $\frac{-b'_0 z^{-3}}{\left(z^{-1}-r\right)^3}$, which can be achieved by minimizing the other three coefficients of the numerator of $H_S(z)$. This implies the use of coefficients b'_i of alternating signs. As a matter of fact, the factor of the numerator in z^{-3} does only delay the signal by three clock periods $T_2=1/f_2$.

With a low frequency sinusoidal state $f << f_2$, we have $z^{-1}=\exp\left[-j\frac{2\pi f}{f_2}\right]\approx 1-j\frac{2\pi f}{f_2}$. Taking into account the expression of r, the factor $(z^{-1}-r)$ raised to the cube in the denominator becomes

$$\approx 1-j\frac{2\pi f}{f_2}-\left(1+\frac{a'_1}{3}\right)=-\left(\frac{a'_1}{3}+j\frac{2\pi f}{f_2}\right)=-\frac{a'_1}{3}\left(1+j\frac{6\pi f}{a'_1 f_2}\right)=-\frac{a'_1}{3}\left(1+j2\pi f\tau\right).$$

The corresponding time constant is thus $\tau=\frac{3}{a'_1 f_2}$. However, filtering will be all the better assured when the time constant is large compared to $1/f_2$, which would require $a'_1 << 1$, which is a condition that contradicts that previously deduced. Consequently, the condition $a'_1 >> 1$, which makes the noise transmittance of the present filter very close to that of a third-order derivator, prevents an efficient low-pass filtering of the signal.

An alternative solution would consist of choosing a digital interpolating filter (see Figure 2.30 with ADC and DAC suppressed) which is capable of implementing low-pass filtering of the signal and high-pass filtering of the noise as well. This latter effect is not really a drawback if the corner frequency is chosen sufficiently large compared to the useful bandwidth at lowest frequencies, meanwhile significantly smaller than f_2, because there is practically no difference between high-pass filtering and derivation in this frequency range, provided the system order remains the same. As a conclusion, if derivation of the noise is the priority, the present filter is a good choice, but if low-pass filtering of the signal is demanded in addition, the purely digital interpolating filter will be a better solution.

Bibliography

[BAI 96] BAILLIEU F., BLANCHARD Y., LOUMEAU P., *Capacités commutées et applications: filtrage, conversion, microsystème*, Dunod, Paris, 1996.

[CD 4046] CD 4046 datasheet acquired from Harris Semiconductors, Texas Instruments Inc., Dallas, USA, revised 2003.

[DAN 02] DANESHGARAN F., LADDOMADA M., "A novel class of decimation filters for delta-sigma A/D converters", *Wireless Communications and Mobile Computing*, vol. 2, pp. 867–882, 2002.

[HC 297] CD74HC297 datasheet acquired from Harris Semiconductors, Texas Instruments Inc., Dallas, USA, 1997, revised 2003.

[JAN 11] JANSSEN E., VAN ROERMUND A., *Look-ahead Based Sigma-Delta Modulation*, Springer, Netherlands, 2011.

[MAT 94] MATLAB® Manual, MATLAB® software, The Mathworks Inc., Natick, USA, 1994–2018.

[MUR 17a] MURET P., *Fundamentals of Electronics 1: Electronic Components and Elementary Functions*, ISTE Ltd, London and John Wiley & Sons, New York, 2017.

[MUR 17b] MURET P., *Fundamentals of Electronics 2: Continuous-time Signals and Systems*, ISTE Ltd, London and John Wiley & Sons, New York, 2017.

[NDJ 16a] NDJOUNTCHE T., *Digital Electronics 1: Combinatorial Logic Circuits,* ISTE Ltd, London and John Wiley & Sons, New York, 2016.

[NDJ 16b] NDJOUNTCHE T., *Digital Electronics 2: Sequential and Arithmetic Logic Circuits*, ISTE Ltd, London and John Wiley & Sons, New York, 2016.

Index

G, H

I

L, M

N, O, P

Q, R, S

V, W

Other titles from

ISTE

in

Electronics Engineering

2018

MURET Pierre
Fundamentals of Electronics 2: Continuous-time Signals and Systems

2017

MURET Pierre
Fundamentals of Electronics 1: Electronic Components and Elementary Functions

BUCCI Davide
Analog Electronics for Measuring Systems

2016

BAUDRAND Henri, TITAOUINE Mohammed, RAVEU Nathalie
The Wave Concept in Electromagnetism and Circuits: Theory and Applications

FANET Hervé
Ultra Low Power Electronics and Adiabatic Solutions

NDJOUNTCHE Tertulien
Digital Electronics 1: Combinational Logic Circuits
Digital Electronics 2: Sequential and Arithmetic Logic Circuits
Digital Electronics 3: Finite-state Machines

2015

DURAFFOURG Laurent, ARCAMONE Julien
Nanoelectromechanical Systems

2014

APPRIOU Alain
Uncertainty Theories and Multisensor Data Fusion

CONSONNI Vincent, FEUILLET Guy
Wide Band Gap Semiconductor Nanowires 1: Low-Dimensionality Effects and Growth
Wide Band Gap Semiconductor Nanowires 2: Heterostructures and Optoelectronic Devices

GAUTIER Jean-Luc
Design of Microwave Active Devices

LACAZE Pierre Camille, LACROIX Jean-Christophe
Non-volatile Memories

TEMPLIER François
OLED Microdisplays: Technology and Applications

THOMAS Jean-Hugh, YAAKOUBI Nourdin
New Sensors and Processing Chain

2013

COSTA François, GAUTIER Cyrille, LABOURE Eric, REVOL Bertrand
Electromagnetic Compatibility in Power Electronics

KORDON Fabrice, HUGUES Jérôme, CANALS Agusti, DOHET Alain
Embedded Systems: Analysis and Modeling with SysML, UML and AADL

LE TIEC Yannick
Chemistry in Microelectronics

2012

BECHERRAWY Tamer
Electromagnetism: Maxwell Equations, Wave Propagation and Emission

LALAUZE René
Chemical Sensors and Biosensors

LE MENN Marc
Instrumentation and Metrology in Oceanography

SAGUET Pierre
Numerical Analysis in Electromagnetics: The TLM Method

2011

ALGANI Catherine, RUMELHARD Christian, BILLABERT Anne-Laure
Microwaves Photonic Links: Components and Circuits

BAUDRANT Annie
Silicon Technologies: Ion Implantation and Thermal Treatment

DEFAY Emmanuel
Integration of Ferroelectric and Piezoelectric Thin Films: Concepts and Applications for Microsystems

DEFAY Emmanuel
Ferroelectric Dielectrics Integrated on Silicon

BESNIER Philippe, DÉMOULIN Bernard
Electromagnetic Reverberation Chambers

LANDIS Stefan
Nano-lithography

2010

LANDIS Stefan
Lithography

PIETTE Bernard
VHF / UHF Filters and Multicouplers

2009

DE SALVO Barbara
Silicon Non-volatile Memories / Paths of Innovation

DECOSTER Didier, HARARI Joseph
Optoelectronic Sensors

FABRY Pierre, FOULETIER Jacques
Chemical and Biological Microsensors / Applications in Fluid Media

GAUTIER Jacques
Physics and Operation of Silicon Devices in Integrated Circuits

MOLITON André
Solid-State Physics for Electronics

PERRET Robert
Power Electronics Semiconductor Devices

SAGUET Pierre
Passive RF Integrated Circuits

2008

CHARRUAU Stéphane
Electromagnetism and Interconnections

2007

RIPKA Pavel, TIPEK Alois
Modern Sensors Handbook

Printed and bound by CPI Group (UK) Ltd, Croydon, CR0 4YY
16/09/2026
14963074-0002